Systematic Mixed-Methods Research for Social Scientists

Wendy Olsen

Systematic Mixed-Methods Research for Social Scientists

Wendy Olsen
Department of Social Statistics
University of Manchester
Manchester, UK

ISBN 978-3-030-93147-6   ISBN 978-3-030-93148-3   (eBook)
https://doi.org/10.1007/978-3-030-93148-3

This Palgrave Macmillan imprint is published by the registered company Springer Nature Switzerland AG.
The registered company address is: Gewerbestrasse 11, 6330 Cham, Switzerland

Preface

Mixed methods is a big area and this book takes a unique approach to it. Mainly working with realist conceptual frameworks, the research designs described here are easier to integrate with theorising and social science, and are less 'pragmatic', compared with some other works.

This book includes the study of ordinal and non-scaled variables, such as ethnicity; patterns across contrasting social groups; multiple cases at different levels; and multilevel interconnections. This book will be helpful for professionals, students, multidisciplinary researchers, and large research teams.

Some of the strengths of this book include: its discussion of strategic structuralism; explaining what realism is and ways to apply it; making explicit contrasts with some other approaches; its description of case-study methods as an option; its suggestions for simple and expanded statistical modelling; and a light development of institutionalism in the application of statistical logic. This book is organised around a series of linked logics of analysis, which can include induction, inference, deduction, retroduction, holism, synthesis, and others. These are introduced as we go along and summarised in Chap. 10.

This book is aimed at any researcher, ranging from beginner to advanced. Researchers can develop wide-ranging, ambitious, international-comparative, and even global projects. These can become large projects, or theory projects. Both such project types typically do need mixed-methods teams.

How to Use This Book

Chaps. 1, 2, 3, 4, 5, 6, 7, and 8 offer guidance for statistical models, like regression, as well as action research, interviews, and other techniques. The first part (Chaps. 1 and 2) is about methodology broadly speaking, and the second part (Chaps. 3, 4, 5, 6, 7, and 8) is about applying the techniques. The third part (Chaps. 9 and 10) covers epistemology (i.e. knowledge claims). Several chapters have an appendix giving illustrations of the specific techniques and the different types of variables or data. An online annex helps further by providing more detailed software-based methods.

This book stresses the ways one can interpret the qualitative data and quantitative data in an integrated way. The chapters explain how a coherent, assertive approach to 'valid' knowledge can be developed. Epistemology is the technical word for this aspect of research findings. Validity and multiple standpoints are central in this book. There are good reasons why we pay attention to the opposing arguments in a debate; we look at evidence in relation to all the arguments. Thus, there can be competing arguments, sometimes associated with key stakeholder groups, and debate can make reference to the internal and external validity of those arguments as well as to their strengths and weaknesses. Evidence, information, and data can be drawn upon to gauge strengths and weaknesses, but social theory itself is also important.

Readers will each develop their own approach, so I have shown here how I would try to develop warranted, robust arguments. I have taken a cautious approach to truth, and therefore this book is compatible with a wide range of positions in social science. In Chap. 9 the idea of a 'grey area' of competing true statements is presented. This approach can be useful if your topic branches into natural sciences, mathematics, artificial intelligence, or another type of science. For instance, it may be that deductive arguments which are 'true' in mathematics do not hold up when applied to a specific, concrete social situation. Additional contributions made by social science and by social theory are highlighted in this book.

Some of the techniques used, which are new and perhaps unfamiliar, are spelt out in the online annex. These include statistical tests for claims about causal mechanisms using case-study research (Chaps. 6 and 7) and routines for keyness and discourse analysis (Chap. 8). Thus quantitative case-study data, statistical data, and textual data are all used in the optional online annex. The software used in the annex includes QSR NVivo, Microsoft Excel, R, and Stata.

Systematic analysis allows you to display data in concise tables or diagrams. Mixed methods could, for example, link a set of semi-structured interviews with a demographic factsheet. As a result, plot diagrams can be presented alongside a discourse analysis. Mixed methods can also include the analysis of secondary survey data combined with textual data. Systematic mixed-methods research (SMMR) projects might also include some large-scale or comparative data analysis.

After studying both systematic and qualitative evidence, one may want to re-visit and revise the original theoretical framework. Using ingenuity to recast, or re-visualise, the research topic is one way to focus and develop a mixed-methods project. In this book I stress structuralist and institutionalist approaches to the real world but you may have other options. The task is not just to know more about the nature of things, but also deliberately to explore the nature of things, and to know how society changes. In books on science, the 'nature of things' is explored using the technical area known as ontology. This book is not heavy on ontology but it does touch on this matter. A further explanation is found in the conceptual part of the book, particularly in Chap. 1.

Ontology is useful when we want to merge a statistical logic with other logics. There are feedback loops from early findings to new data and back to findings. By putting forward this approach, this book is unique because it rejects the usual approach that stresses a 'schism' of qualitative and quantitative methods. In addition, using the feedback loop approach, you can also widen the arena of research. One can encourage reflexivity during a project, and/or create a Rogerian debate. A Rogerian debate is a two-argument contrastive debate, following the concept put forward by Carl Rogers that we can respect all the arguments of different agents. There are several possible ways to calibrate variables and variates; and fuzzy sets might increase your sensitivity to different kinds of causation. Overall a mixed-methods project can use a wide range of methods of logic and these are each explained in Part II, that is Chaps. 3, 4, 5, 6, 7, and 8.

This book avoids a common trap known as 'atomism'. In statistics, atomism might mean assuming all the units in a population are fundamentally identical in some ways, for example the units could be persons, or firms, or farms. Atomism is a simplification in which a flat model of the world is used. Actually, the world itself is multilevel. Therefore, atomism is widely recognised to be problematic. Business studies, for example, would err if it treated all businesses like atoms. This book stresses the real diversity of the units in Chaps. 4, 5, 6, and 7. It can be useful also to apply reflexivity and work up a detailed social ontology (Chap. 8), re-working what we know to re-describe the entities in the society. All these methods are useful and can be combined judiciously when one is planning a research project.

Another method of analysis is to apply standard regression or bar charts, involving a typical logic of hypothesis-testing or inference. It is not necessary to fall into the trap of thinking that only inference or only falsification can establish knowledge. I explain inference, falsification, and **retroduction** (Chaps. 1 and 2), and then show how to establish new findings, based on evidence (Chaps. 3 and 4). Retroduction means adding new evidence to your basket, which leads towards new findings. Some people say retroduction is 'asking why', but here we also seek the evidence to answer those 'Why' questions. Your retroductive moves can use statistical or qualitative evidence or both. I argue in favour of several systematic methods of gathering and using data—fuzzy sets, binary measurements, coding qualitative texts, and others. Overall this book is consistent with multidisciplinary and holistic social science.

In Chaps. 9 and 10, I set out a challenge for researchers based on my belief that our norms do influence our research. I explain in Chap. 9 that we are reaching situated knowledge, not objectively true or permanently true knowledge. Taking the 'situated knowledge' approach strengthens your ability to engage in meaningful debates. These could be explanatory debates or policy arguments. They would be heavily rooted in evidence and the debating parties would cite their information sources. Readers of those debates gain an ability to develop solid, well-built arguments. In Chap. 10, I summarise and bring together the main arguments of this book.

Manchester, UK   Wendy Olsen

Acknowledgements

I thank the many people who helped me and contributed to this book, especially Jamie Morgan and David Byrne.

From 2008 through 2019 I held workshops on mixed-methods research alongside numerous co-organisers including David Byrne, Steph Thomson, and Patricio Troncoso, in the United Kingdom, India, Germany, Bangladesh, Japan, and other places. Jointly with Steph Thomson and Lina Khraise, I held summer schools in Manchester on mixed methods. We had support from the National Centre for Research Methods to teach fuzzy sets.

I am grateful to the British Academy, the Japanese funding agency JSPS, the Economic and Social Research Council (ESRC), the UK Department for International Development (DFID), the University of Manchester Faculty of Humanities and School of Social Sciences, the Cathie Marsh Institute, Department of Social Statistics, Methods@Manchester, and the Alliance Manchester Business School in particular. They gave valuable support and financial inputs.

I especially thank the University of Manchester for seed-corn grants. We have a supportive research environment. I am very grateful to Jihye Kim for her ongoing support and mixed-methods teamwork.

I thank the British Academy particularly for granting our team funds to study Innovation in Global Labour Research Using Deep Linkage and Mixed Methods, grant number PM140147 2014-7.

I thank the DFID and ESRC for the grant on Gender Norms, Labour Supply and Poverty Reduction in Comparative Context: Evidence from Rural India and Bangladesh. Funding from the Economic and Social Research Council and DFID was under grant number ES/L005646/1, 2014–2018.

I am very grateful that in 2020 the Global Challenges Research Fund (GCRF) gave my team funds to study Social-Action Messages to Reduce Transmission of COVID-19 in North India. I led this as principal investigator, along with, Dr. Arek Wiśniowski, Prof Amaresh Dubey, Dr. Purva Yadav, Dr. Manasi Bera, and others. The GCRF funds arose within Research England, a UK public-sector funding source. In 2019 the Global Challenges Research Fund had also given our team funds to study Charters for Better Work Better Lives: An Indian Partner Network. For these two mixed-methods studies, we are very grateful. I particularly appreciate my team members and co-investigators, notably Dr. Jihye Kim, who helped me grasp how to develop the research design and interpret the results without going back to the usual 'schism' of qualitative versus quantitative methods. Not all the GCRF results are represented in this book, but I have tried to depict and explain some of the wisdom we developed.

Contents

Abbreviations

ANOVA	Analysis of variance
CMD	Common mental disorders
CMO	Context-mechanism-outcome approach, rooted in complexity theory
FSGOF	Fuzzy-set goodness-of-fit test, an F-test described in the online annex of this book
fsQCA	Fuzzy-set QCA, also refers to software by that name
GHQ	General Health Questionnaire
ICC	Intra-class correlation coefficient (closely similar to variance partition coefficient VPC)
MCA	Multiple correspondence analysis
MVQCA	Multi-valued QCA
NVivo	A specific software package for analysing qualitative evidence (brand name), sold by QSR (a company)
QCA	Qualitative comparative analysis
R	A specific software package for many statistical, matrix, graphics, and other functions (brand name; open access)
ROWE	Results-only workplace environment
S I M E	Structural-institutional-mechanisms-event framework
SMMR	Systematic mixed-methods research
STATA	A specific software package for many statistical, matrix, graphics, and other functions (brand name)
TMSA	Transformational structure-agency framework
TOSMANA	A specific software package for analysing comparative case-study evidence including making boxed Venn diagrams (brand name; open access)
VPC	Variance partition coefficient, one measure of the variance decomposition in a multilevel model

List of Figures

List of Tables

Introduction

This book offers new ways to do mixed-methods research using quantitative methods. This book helps in situations where one may not want to carry out a statistical analysis but one does want (or need) to do a systematic analysis. It also shows connections between statistical analysis and social-science argumentation. This book uses the term 'systematic methods' to refer to the many modes of social research that involve quantification. These methods can be combined with other methods, creating complex mixed-methods designs at the planning stage. The term 'systematic' also refers to using categories that make explicit the contrasts across cases. Systematic evidence often involves quantified measurements, but it is also dependent on theoretical frameworks. An obvious example is the Yes/No variable known as a 'dummy variable', and ordinal variables are another example.

The methods chosen in a project bring the researchers into contact with social reality, generating evidence. There are multiple layers of meaning, and multiple types of entities, that each project will tap into. The research team will want to understand some of these in its attempt to make sense of the world. This book helps you gather, harvest, and interpret both quantitative and qualitative data, and bring a project to a successful conclusion.

Overall, this book helps you use mixed-methods approaches with systematic data and increase your sense of confidence about the resulting claims.

Part I

Setting Up Systematic Mixed Methods Research (SMMR)

Mixed Methods for Research on Open Systems

1

Researchers can use a wide variety of mixtures of methods. We can mix data types and mix logics of interpretation. This chapter aims to present some options among all these possibilities. I am including quantitative data and algebraic representations, thus focusing on more systematic approaches. I also present two broad approaches to social science research, which can be applied very widely:

- Systematic Mixed-Methods Research
- Realist Action Research

These broad approaches are tractable as models to follow. You will find overlap here with other textbooks, too; many other options also exist. Each researcher may have their own philosophical basis, which may not match the ones used here. Philosophies also change and develop over time, both at the personal level and in research teams. Therefore, textbooks on mixed-methods research might not always be consistent with your own approach at a fundamental level. This is the methodological level. The aim here is to increase your sensitivity to several of the key issues in these debates. In this chapter I focus mainly on quantification, triangulation, and realism. Later chapters take up other issues.

Starting with quantification, note that two projects could use similar methods or carry out the same algebra, while yet being fundamentally different at the level of theory, explanation, or methodology. It helps to distinguish 'methods' from 'methodology'. Two projects could use the same method—such as a regression—for different reasons, in a different context. Authors need to spell out a little of their methodology and particularly some of the key 'premises' of their main argument. The audience will appreciate your depth and clarity. Elements of your methodology include:

- which disciplines you are making reference to;
- what initial assumptions are made about research scope;
- what kind of holistic and atomistic units are being studied, and how;

© The Author(s), under exclusive license to Springer Nature Switzerland AG 2022
W. Olsen, *Systematic Mixed-Methods Research for Social Scientists*,
https://doi.org/10.1007/978-3-030-93148-3_1

- what is the scope of the research, and to what range of phenomena would any generalisation apply;
- how the selection of cases or sampling and measurement were carried out, and why;
- what evidence was recorded and how transparency is achieved in the study;
- what is the basis for the validity of key findings;

For quantitative studies some helpful guides are found in Goldthorpe (2001) which focuses on causation, and in Deaton and Cartwright (2018) which focuses on validity. Some concepts above may be unfamiliar, so a broad introduction to concepts is found later in this chapter. Holism and atomistic entities are covered in Chap. 8 of this book. In Chaps. 3 and 4, I spell out key issues for interpreting statistical results. Concerns about measurement error may be focused on by reading Chap. 5.

In some projects, it may not be necessary to go into detail about causality and epistemology but instead one may follow an exemplar and cite that exemplar as broadly the source of the methodology. For instance, a systematic review could cite the key works on foundations of systematic review. However, even if deep issues of methodology are left implicit, every research report should present its detailed research design. The design section needs to include both the rationale and details of methods used. This is widely known when the research involves fresh surveys and interviews. It is also useful when reporting on action research, because the readers need an account of the stakeholders and participants. These steps are also useful markers when we review, or plan, a research career based on mixed methods. The **mixed-methods research career** overall is bigger than a single research project.

We can now look in more detail at what quantification achieves (Sect. 1.1). A conceptual glossary is offered in Sect. 1.2, aimed at avoiding a supposed 'schism', or contradiction, between quantitative and qualitative research methods. There is then a section on triangulation with examples (Sect. 1.3) and a further explanation of key concepts used in discussing mixed-methods methodologies (Sect. 1.4).

1.1   The Link Between Quantification and Mixed Methods

The quantification step in a research design involves developing a representation. There is always an agent, a voice, and an audience. The agent is the researcher, their voice is emergent over time (and may be a multi-author voice); and audiences are several. In structure-agency terms, 'to quantify' is a discursive act of an agent and this act has—at first glance—very little structural determination behind it. Yet the voice that emerges is evidence based, and the evidence may be structured. Therefore, researchers have plenty of scope to argue in favour of quantifying something, such as marriages. Quantification also involves harmonisation where a symbol is made to 'represent' something complex, like a social entity, across a broad scope. We could measure emotional intelligence in a classroom, and the scope is 'this classroom now'. Emotional intelligence refers to a range of social skills (see Goleman 2004).

A controversy could arise when someone points out, rightly, that the quantified 'thing' had features that go far beyond those found in the numbers. For example, social norms of emotion management would underpin the whole class's response to a survey on emotional intelligence. We could measure the social-emotional skills, or we could measure the conformity of students with social norms about social-emotional skills. Quantified measures generally represent things by a correspondence. The 'thing' the measure indicates usually cannot exhaust or represent the whole of the phenomenon. Instead, we create a measure to reflect some essential feature(s). Thus personal social-emotional skills might be our focus. A different study might look at what students think are the norms about social-emotional skills—which is a different focus. In setting up quantification, researchers try to express ourselves clearly. We arrive at a resolution of arguments about how much heterogeneity there is. (More on this in Chap. 5.) Things we measure can be both personal and social at the same time. As seen in Chaps. 7 and 8, the measurements reflect both micro and macro elements at the same time. The children in one classroom could be diverse in social backgrounds, yet have some key shared characteristics. Measurement can reveal both (Chap. 5).

Besides information on quantums, values too are embedded in many measures (Olsen 2007). These embedded values are both the values of the agent and the social norms of the surrounding society. Measuring well-being offers an obvious example (Neff and Olsen 2007). The act of choosing what constitutes well-being reflects both personal and social values. Usually the socially general values are called norms. I return to issues of value and norms in several places in this book. In the complex modern world, values often collide, so we need an encompassing overview as a background for doing scientific research. Plurality of values is part of how we define 'modern'; contrasted with 'traditional' which is a term often used to try to refer to more coherent or homogeneous social forms. When researchers are open to a wide range of value systems, 'Systematic Mixed-Methods Research' (SMMR) could be a way to feed evidence and claims, based on clear arguments, to the inevitably diverse audience. When we try to invoke, develop, and refine our team's opinions, and in the process allow ourselves to be transformed and to work on transforming society, we are moving towards 'Realist Action Research'. These distinctions will be summarised and clarified further on, particularly under the heading of reflexivity. Many other mixed-methods possibilities also exist.

My focus will be to show how quantitative methods weave in and out of both these approaches to research. Molina-Azorín said 'mixed methods research requires that scholars develop a broader set of skills that span both the quantitative and the qualitative' (2011: 19). I will present a broad-based view that 'quantification' belongs in systematic mixed-methods research. Using tabular data or numbers is helpful in action research, too. This book explains how all these various activities and their underlying rationales fit together.

Overall it is a useful book for those doing triangulation (using data gained from three or more distinct vantage points). The competing metaphorical vantage points can be socially different, or even physically alternative, modes of data collection. For example interviews, survey data, and participant observation would be three data types, and hence triangulation. Or they might be competing voices from agents in a scene. Another form of triangulation is to use government data, interviews, and workplace shadowing. Many other mixtures are possible.

We distinguish three main levels of quantified measurement (categorical, ordinal, and continuous) at an early stage of handling quantitative data. But measurement itself has three kinds, which refer to different dimensions of quantification. The first is when we assign an algebraic parameter to a thing, for example sex is a variable α (*alpha*). The second is when we assign numbers to different categories for a given entity, such as the type of marriage. The third is measuring on a scale or continuum.

One always has to clarify the 'thing' being defined, for example one might state that marriage is a lasting relationship with social sanction, entered into voluntarily by two people. Then the indicator gives values which name the options in a specified set, for example: {married in a church; married with a registration ceremony but not in a church; married in a synagogue; married in a Hindu ceremony; otherwise married}, numbered {1, 2, 3, 4, … 99}. Thirdly, we can also quantify by making numerical estimates of relationships, and there are many alternatives for this. One simple example might be the different heights of bars in a bar chart: each reflects 'the mean' of a variable Y for a given value of variable X. The relationship here is that of categories of X and the grouped amounts of Y. A more complex example is the regression slope parameter (Chaps. 3 and 4). This book takes up issues of quantification in relation to mixed-methods research designs.

In summary, we quantify by algebraic abstraction, by labelling types and then counting occurrences, or by assigning a number to a relationship. These can be combined.

Many other forms of quantification, such as using probabilities, also exist. The controversy attached to the labelling of types arises because we must either involve ourselves in theorising, or else engage with lay discourses, to arrive at the labels for each type of marriage. Theorising refers to assigning labels to an array of things, processes, acts represented by verbs, and so on, till you can put these elements together into a sensible description or explanation. There is no avoiding the 'theorising' step in research. Without it, your research will be stale. You stipulate what things you will discuss and what their types and typical features are. In mixed methods, we have a supplemental task. We engage with lay discourses, allowing ordinary people in society to show you their way of theorising and making sense of the world. 'Lay' discourses here refer to discussions that use ordinary everyday wording (see Sayer 1992). The lay person is highly expert about life-world domains, and they can tell you a lot by sharing their descriptions with you. Instead of merely adopting their language, you develop your own encompassing mode to describe things. We may use a meta-discourse (meta meaning 'above' or 'outside'). You write: 'They said that "… XXX … occurs" to describe this phenomenon, which we call YYY'. Scientific discourse can thus be both expert and complex. It can be either reflexive or unreflexive (see next section). The complexity of competing discourses in social reality matters very much for the choice of research methods.

To settle down controversies, a thoroughly researched theoretical framework is helpful. In development contexts where global trends and contrasts are the focus, research teams are guided in part by funding agencies such as Department for International Development (DFID; see UK Aid Connect 2018). They recommend formulating a 'theory of change' and then using this, deriving the strategy for causal analysis. The theory of change approach involves a discussion around how narrow/wide the net is to be cast for a project (see Funnell and Rogers 2011; HM Treasury 2020; UNDP/Hivos 2011, cited in UK Aid Connect 2018). For DFID, causality is important. Other options also exist, though.

Other theoretical frameworks achieve a synthesis of literature. For example, we could propose:

– A feminist approach to styles of teaching in schools in the UK.
– A psychodynamic and sociological approach to drug-taking patterns among students in England.
– A socio-economic study of work at home and for pay in rural South Asia (Chap. 8).

All three of these projects imply a theoretical framework, and after doing research we may develop this framework more. Numerous kinds of theories are workable. I like those which have a depth ontology and a realist approach to representation. Another kind of theorising invokes a strong social constructionist interpretation; and a third might be a grounded theory about the phenomenon. Among these three, the first lends itself best to the use of quantitative methods. To review how to theorise and where theories sit within your research design, see Bryman (1996), Layder (1993, 1998), Blaikie (1993, 2003), Blaikie and Priest (2013), and Outhwaite (1987).

There are many ways to approach mixed-methods research. One might want to represent multiple voices in your research or analyse a set of distinctive narratives whilst recognising that your own text becomes a new, synthesising narrative. This book will help in all these cases. We can even quantify types of narratives or make a typology of discourses. In this book I explore how quantitative methods can help us to discern a dominant discourse, and I show how essential it is to be able to discern in a more qualitative way what resistance has occurred, and what discursive challenges are occurring. This kind of enquiry distinguishes 'social statistics' from statistics itself, but also opens up a social scene for any research project. We will want to notice what deviations are happening in the non-dominant discourses, and how and why it is so. A deviation is a surprising turn of speech. In labour markets for example, while government is talking about regulations and minimum wages, the non-dominant discourses might focus upon getting respect by being argumentative, or may be planning sabotage or trying to end zero-hours contracts. To discern these patterns one needs larger data sets, not just small-scale ethnography. Mixing quantitative with qualitative methods would nearly always prove rich and worthwhile.

I will set out the depth ontology approach later in this chapter, making explicit my preferred realist angle on it. I have argued in favour of critical realism of a particular kind elsewhere (Olsen and Morgan 2005). The depth-ontology realist approach has won numerous arguments. For example, we challenge the use of excessively individualistic frameworks (see Lewis 2000). In each project, this approach needs some fresh development to make applied research work well. Realists urge us to apply a depth ontology. For instance a study of marriage in India would have to be well grounded in local conditions, well aware of the existing research, and able to use relevant languages and dialects. Religion and law would be relevant. I am interested in the pros and cons of 'using', or even creating, empirical data in mixed-methods projects, including those which are international and comparative, so I devote space to the depth ontology issue throughout the book.

Many scholars resist 'realism' itself and many are not attracted by 'critical realism', partly because these two frameworks are so large, difficult to grasp, or have a negative image, publicly displayed by their critics. Yet these schools of thought each have many diverse scholars attached to them, and one wants to consider the issues rather than joining a school of thought. In addition, the two schools can seem contradictory (see Further Reading). In addition, each researcher wants to set out their own methodological framework. I aim to present some useful tenets and principles, but still keep this text open to many other interpretations and approaches. Comments are welcome, as my own perspective changes with time as I learn more.

This book offers research designs for several kinds of mixed-methods research projects that include quantified data. I will now proceed to introduce triangulation and some key terms: open systems, retroduction, realism, and the integration of methods. The key point is that you can mix up the various kinds of methods in one project. Later in this book I will also discuss the role of science in society, and you as a scientist making your way in a mixed-methods research career.

1.2   A Conceptual Introduction to Methodology and Ontology

▶   **Tip** The validity of the views of stakeholders whose voices are heard in a project needs to get fair consideration. Some views may be externally valid but others may have a factually invalid element, and yet each particular view or belief may be coherent from within that stakeholder's world-view. The research task includes getting an understanding of these world-views. In some literature, these world-views are also called standpoints.

A project needs the right methods for the topic and the research question that you have carefully worked out. The chosen topic will dominate your decisions perhaps much more than your own initial philosophical starting-point. Being sensitive to the interplay of your initial prejudices and your evolving findings is known as being 'reflexive'. A reflexive person is thoughtful and is aware of change in themselves. Keeping a research diary helps you to record issues and new theorising, and it also helps people realise which old thoughts are being challenged.

The Glossary of this book covers some of the key concepts needed to deepen the epistemic basis of a quantitative or SMMR project. For example, for the term 'reflexive', a definition is offered: '**Reflexive**. The reflexive methodologist allows for their own introspection, and dialogues among research participants, as a fundamental part of the learning that goes on during social research. In this context, reflexive means reflecting upon one's own approaches, one's language and body responses, and others' language, and moving toward new images of the scene'. In the rest of this section, I will set out working definitions of 'realism', 'methodology', 'modern' and 'late modern' approaches to research, and reflexivity. Then I will give an example of how reflexivity can affect a project from the start.

Realist research refers to the real world which underpins and precedes the specific project. It places the knower(s) as a creator of situated knowledge (Smith 1998). The knowledge-creation process is scientific, but the knowledge claims are not person-specific. (Their validity is not just a subjective matter; see Letherby, Williams and Scott 2012.) Indeed the results of a mixed-methods project can be just as reliable as the results of other kinds of projects. The key feature is to recognise that the scope of a given project can be limited, and that abstract results may draw out lessons for particular types of social entity. Thus reliability, which is where the next researcher can validate and rediscover the findings that you drew out of your project, can be a feature even in a rapidly changing social world with competing social narratives.

I also advocate a late-modern approach to research (Chouliaraki and Fairclough 1999). The use of empirical data, transcripts, quantitative data, algebra, and surveys has aspects of modernity, including harmonisation, a large scope, scientific sophistication, and an impersonal approach to some aspects of validity. However systemic mixed-methods research also has a 'late modern' quality, for two specific reasons.

1. Late modern research admits the situatedness of each research project. We realise the world is composed of many different historically specific trajectories, refractions of social groups which are not fixed identities, nor are totally separate, but which have a shared history.
2. Late modern research admits the deep complexity of our world, and this change of focus widens the globalised scope of research. Complexity leads to fundamental reflexivity and social reflexivity.

Using the late modern approach we can generate better, more valid, less ethnocentric findings.

The active transformative approach (recommended by Flyvbjerg 2011) would be more prominent in 'Realist Action Research' and less prominent in 'Systematic Mixed-Methods Research'. The more critical approaches to social research take risks and invoke values in ways avoided by the supposedly 'value-neutral' methods (Olsen 2010, i.e. Chapter 1 in Olsen, ed. 2010). The pretence of having a value-neutral science has led a wide array of researchers to claim they are doing one thing (creating knowledge without applying their own personal values) when actually they are doing another thing (invoking their value system to strategically aim to transform society through some kind of future action). For example randomised control trials when applied in social contexts were once widely considered to be a mono-method quantitative method, but the stance has been revised towards mixed methods by several authors (Olsen 2019; Masset and Gelli 2013; Deaton and Cartwright 2018). The shift to mixed methods has many proponents, who argue their case differently. To give just two examples, some applied Randomised Control Trial researchers argue that focus groups help the researchers develop insight and see linkages (Masset and Gelli, op cit); while a broad range of social scientists argue over whether we should be looking for the causes of effects, or the effects of the causal mechanisms (Deaton and Cartwright, op cit.; Bonell et al. 2018). What is meant by this is important: those looking for the causes of effects have a focus on the dependent variable, and its causes, and they keep this focus clear within a project. However those who are interested in the effects of causes are willing to engage with the non-extant situations, and why some counterfactuals simply could not occur. Thus, for example perhaps if a holistic cause is absent in Country A, then Treatment T causing outcome Y to arise in countries B, C, and D could not be correctly generalised to apply in country A. One group might omit A from consideration. The others would ask why A is different, and thus by having an interest in causal mechanisms, go on tracing causes backward.

For some in the statistical community this is known as infinite causal regress, because logically the deep or distal causes cannot all be enumerated. But for those in some competing sociological and theory-based communities, researchers may feel intrigued and may wish to pursue some of these historical and macro avenues.

Yet the problem with looking at too many things is that the concrete focus often narrows to a single locality, and then two typical dilemmas of 'observational studies' arise. These were enumerated by Concato and Horwitz (2018: 33): first, in an observational study where 'treatments' are not controlled and random sampling is not used, the data may not be balanced across key features, leading to misguided interpretations; and secondly sometimes data quality is compromised by the lack of a clear, focused 'Research Question'. Overall, mixed methods research has grown in popularity due to the wide acceptance of both strengths and weaknesses of the competing types of research designs.

The notion of synthesis via meta-analysis, or via systematic review (which is almost intrinsically multiple-methods based), was discussed by Duvendack et al. (2012). They argue that the complexity of combining studies leads to valuable knowledge in spite of the inevitable expensiveness of the procedures overall.

It is worthwhile pondering these aspects from time to time because research and its role in society change over time. In some countries research is rather separate from policy-making in government, but it still affects third-sector and private-sector practical actors. In other countries, like those in Europe and the OECD, a complex set of research institutions within and beyond government influences what is considered good-quality research both outside and inside government. The 'beyond' government part includes think-tanks, foundations, learned societies, and other organisations. Researchers using mixed methods can influence the structures and norms of research. Instead of being passive, we can be active agents of how social research is perceived. We can influence what role research plays. Action research is a moment of active engagement with an organisation or a 'division' or partner, lasting from about 1 month to 10 months or so. In some situations, you cannot carry out action research. Creating opportunities for action research involves becoming an expert, developing contacts, cultivating partnerships, and recognising that some standard norms about research quality might not apply during action research (see also Chap. 9). One can combine action research with other projects that involve simply secondary data analysis or mainly primary data collection. Action research experiences build up as part of strong, well-founded mixed-methods careers. The career will tend be both strategic and 'phronetic', as described by Flyvbjerg (2011), with a mixing of researcher-activists and practitioner-activists. One of Flyvbjerg's points was that simply 'describing' a situation may not be a very thorough form of research. Another point is that people intrinsically tend to want to change institutions. Through these action-research experiences, people can aim towards being an influential social actor (Flyvbjerg 2011).

To be specific: for example a study of marriages could be reframed once we began to study the ethnic and religious groups in our chosen location, let us say the United Kingdom. First a variable for length of marriage might be continuous in its level of measurement (data type 1, continuous quantitative). Secondly a range of interview texts can be looked at in relation to marriage (textual data and experiences, data type 2). Thirdly, we may start to classify additional harmonised variables across a group of cases:

> Marriage type is one or more of the following: {married in synagogue; married with presence of imam from a mosque; married through Hindu rites; married with other south Asian rites e.g. Sikh, Jain, or Buddhist; married through Christian church rites; married with registration ceremony but not in a church; other marriage type} numbered {1, 2, 3, 4, 5, 6, 99}.

Without these adaptations, the original variable was ethnocentric and it would also gradually get out of date. An alternative is to make three variables which may better represent this overall situation:

> Variable M: {married or not married, legally, at present}
> Variable C: {married in a home; married in a building of worship; married in another type of place} or not married.
> Variable R: {married by legal registration; not married by legal registration}

The variable M is a binary; C is multicategorical; and R is binary. These measures can be considered both quantitative (i.e. leading to percentage reports) and qualitative at the same time. I will call the qualitative variables 'data type 3'.

This simple example might seem trivial. Yet in the past, studies of 'the labour market' in countries like the USA used only data on males. Yet not only are women very active in labour markets, but also the invisibilisation of their work, calling it a 'private' matter, had started many centuries earlier. Women's inputs to production and services were hidden through patriarchal discourses for millennia. Thus social changes affect the research agenda. Research can also affect and cause social change, though. So even a small measurement decision may turn out to be important in the future. In this section, I have moved from saying what triangulation is towards making it appear likely that the researchers will use triangulated studies to influence society.

In later chapters I discuss measurement decisions in triangulated projects. All quantified data rest upon theoretical frameworks, and hence conceptual representation decisions. Thus they rest upon qualitative exploration of our social world.

Originality in setting up some variables or types is a useful way to achieve innovation in your research. When society is geared to reproducing its key problems, research innovation can help with making change happen.

I will now stipulate my working definitions of two phrases which may be helpful when planning a mixed-methods project: methodology and methods.

Key Terminology

Methodology is a set of assumptions and procedures for research. Yet methodology is distinct from methods. **Methods** are just the tasks or sets of procedures in the research. Research design is an explicit summary of one whole project plan with its stages, dates, intended data types, and outcomes. The research design usually links up how the methodology will underpin the activities that help us move from data to conclusions. All these definitions are independent of the use of the concept of 'ontology', which refers to the nature of what we are studying.

Methodology involves a whole series of key assumption decisions. One may assume the research is limited to the banking sector, or that the project should last only 6 weeks so you cannot use new primary data. These decisions about place and time are important and can be made explicit.

Further methodological assumptions tailor the use of the specific methods you may choose. For example if you aim for induction as your desired way to gain knowledge, then you may decide to gather lots of data from two or more types of source. On the other hand if you assume that you must test every distinctive hypothesis in a theory—or as many as you can—using appropriate data, then you might operate very differently when you reach the methods stage.

The methods stage is a set of actions, often telescoped into certain predictable types of routines. 'Content analysis', 'grounded theory', and 'regression' are examples of methods. They follow routines.

These routines can be called protocols or templates. A protocol is a step by step guide to a method of research. For example randomised controlled trials have about four main protocols: blind, double-blind, multiple control, and quasi-experimental (in situ). A template is used sometimes in qualitative coding of textual and image data. Here a certain form of coding is expected, for example in thematic analysis there is a grid; and in grounded theory there are first retrieval codes and secondly axial codes. Methods include the details of these protocols. Methods can include a plan of how to carry out a semi-structured interview, how to design a questionnaire, or how to convert hospital case records (or legal case documents) into a spreadsheet summary format.

The details of methods protocols are widely known and easily accessible. It is a little harder to get your methodology straight. At least three issues must be dealt with—ontology, epistemology, and validity. There is scope later to deal with these in more detail, so here I simply define them.

Ontology refers to the nature of what is being learned about. Think about Freud's famous theory of the human psyche. He claimed that its ontic nature was trifold: id, ego, and super-ego. He did much more than this, building up a detailed 'ontology' as a study of what the human mind is, and what constitutes it. An ontology is a set of claims and their implicit subtextual content. We may question Freud's theory. In such a case we may revisit psychological data to develop a different ontology; thus an ontology is a theorisation.

Individualism is another example of an ontology. Another would be a racist mindset that distinguishes each ethnic group as distinct endogamous social groups. Thus, even when you have an ontology, your claims are not necessarily ethical, true, or valid.

The word ontic is the root word for ontology. The ontic nature of things is the essential aspect or the core of them. This word 'ontic' helps us express the possibility of misinformation. A thing might have its ontic nature hidden, and evidence about it might be mistaken. There are lies, measurement mistakes, and biases of language; there is also a selection problem when we focus down onto selected types of marriages. So it is obvious that not all ontological claims are necessarily valid or true. In scientific realism, whilst we refer to a thing's true ontic nature, we do it mainly in order to point out the fallibility of human knowledge. There are specific problems, which can be addressed. Thus knowledge can grow towards accuracy by being rooted in an orientation towards reality.

Epistemology refers to the value of the knowledge you have. There are some typical epistemological claims found in textbooks, and some of these clash with each other. Experts suggest that we need to be able to see the world with a coherent approach to knowledge. Thus if all 'voices' are genuine, and all are true, then an inconsistency can arise, which is troubling to many scientific researchers. It may be better to argue that all voices make claims which are genuinely arising in their own social and agentive situation. Then when inconsistencies arise, we don't assume all the claims were true but that all were intended as true. This illustrates how sophisticated wording can help solve epistemological (knowledge) problems. Epistemology has endless variations and invokes quite a few value precepts. An epistemological stance and an ontology together make up part of your methodology. I don't mean these as abstract things; they are very concrete.

For example:

Epistemology: I respect the statements made by the people in the survey.

Ontology: The marriage types are distinct and mutually exclusive.[1]

Another issue is that some people do not respect the knowledge of other groups. A lawyer could, for example, not give high respect to a victim's approach to law. A questionnaire specialist may not trust the words that emerge from a semi-structured interview. These fractures over the **validity** of claims rooted in real-world experience are typical of human society in this century. We mean by 'valid' either that others accept that your claims are applicable and true over a wide range (external validity),

---

[1] Please note that types in a typology do not have to be mutually exclusive. I have put this statement as an argument, which would be made in a particular context. If types overlap, the typology is likely to benefit from fuzzy-set measurement. A simple example are taxi drivers, who may have two or more work statuses at the same time, for example Uber and waged.

or that the claims are applicable and true with reference to the instances to which you refer (internal validity). In past centuries many societies worked by stricter rules so that official knowledge was authorised by a monarch, and so perhaps legitimate knowledge even had to be made as a statement by a monarch, or by certain government outlets representing the monarch. Other forms of knowledge were not acknowledged as valid or worthy. Farmers' knowledge, for example, has widely been considered 'vernacular', or local and inferior, with a negative connotation. Gradually, with the expansion of scientific knowledge, experimental method, and data collection, we have moved to a concept of knowledge that allows for many claims being fundamentally contestable. People widely know that supposed 'facts' may have a poor foundation. We know that we have to provide a justification for our key claims.

In the current period, we find a multitude of claims competing in a rich social space. There are many languages; there are specialist vocabularies; regional dialects; and multiple forms of expertise. All knowledge is socially and historically situated. It is harder now to make strong claims about what is a valid argument, versus what is a strongly believed but not valid argument.

For Example:

Epistemology: Marriage categorisations varied *in situ*, and included some cohabiting pairs, but our expert view is somewhat more abstract and has been simplified here.

Ontology: The cohabiting couple scene was much more varied in its norms than the marriage norms were among those couples who had had a marriage ceremony.

While reviewing triangulation and the key terminology, I have provided examples of three types of reflexivity within mixed-methods strategies. These were personal reflection, competing voices, and the process of expert synthesis. I also showed how the research design can have tacit elements and assumptions. The need for reflexivity is still present when we use quantitative methods, just as much as when we don't use them.

I have argued that all quantitative steps in research rest upon a firm qualitative footing. Yet it is not necessary for qualitative research data to rest upon a firm quantitative footing. The three data types fit together very well. Table 1.1 summarises how a triangulated study can use different types of variables or other measures such as indicators.

Table 1.1 Dimensions of quantification

Data type 1: Representing a quantum via algebra
    Example: Years married is denoted by y
Data type 2: Textual data and visual representations, not represented with variables, but arising from primary or secondary data experiences ranging from interviews to archival research. An example is in Fig. 4.6 where cases are named on a two-way quantified schema
Data type 3: Labelling distinct things using labels for real types
    Example: Types of marriages are denoted by 1, 2, ... 6, 99. Each has a distinct label
    We could call the list of distinct labels a set M containing {…}
A fourth type of number is often used in algebra: Assigning numbers to relationships between entities
    Examples: The correlation of y and x; or the association of y with types of things in the set M

In Table 1.2 I give an example of a **variate** of each of these kinds. Byrne (2002) used the term **variate** to refer to measures about a real sample. He restricted the use of the term 'variable' to situations where we have moved into an algebraic discourse and can talk about a variety of real possibilities and not only specific concrete people. Byrne (2002: *passim*) contrasts variables with the variate traces. Variates are data referring to concrete cases and conditions, while variables refer to real situations, both actual and potential.

The marriage example suggests that measurements indirectly reflect the observing people's religion or ethnic group of origin as well as the features of the marriages themselves. This theoretical thought may lead to improved measurement.

In such ways, reflexivity is helpful in improving quantification, and this is discussed more in the next section which handles reflexivity within triangulated studies. The rest of this chapter is then focused on explaining more about open systems as part of a depth ontology. The material in these sections widens the coverage and remit of 'realist research'. Open systems and a depth ontology are essential points of departure for mixed-methods research. Some readers may not be comfortable with 'realism' or 'critical realism', but the issues are discussed here in ways that may allow points of dialogue to be brought out. Notably, in the reflexivity section, we consider the agent who brings out a representation after a research experience. Then later in the 'open systems' section we again see the agents of social action as being involved in the creation of knowledge. Overall, I tend to stress, we end up arriving at knowledge about the world **in which we are living**, and not knowledge that is from any superior, perfect, or impersonal perspective. By explaining elements of my argument, I hope to make clear why I do consider my work 'realist' but also why mixed methods is a good idea, whether or not the researchers take a realist view.

An Illustration of Reflexivity About a Bar Chart

People are reflexive actors while they are doing statistics or making bar charts from survey data. Here reflexive means thoughtful, self-considering, or working in a reflective manner. We make strategic decisions, invoking our values, and we ourselves reflect and invoke older (prior and ongoing) structural features of the society. We are not value-neutral actors while we do the triangulation and statistical parts of a piece of research. There is scope for differences of opinion about how to interpret data—or even what data to use. Thus, although we are scientists operating in a group, we are not able to reach a perfect consensus on any particular matter. This is normal. There will be multiple voices.

Using data from India I will illustrate the issues at stake, looking at gender, common mental disorders (CMD), and husbands' activities. The authors of the research exerted reflexivity. Although the figures relate to groups of women, the important claim being made is that social factors contribute to individual suffering. Thus although reflexivity might seem personal, science conversations involve social reflexivity, too.

First consider some evidence of social-class differences in the tendency to depression in India (Fig. 1.1).

Shidhaya and Patel (2010) argued that a mental disorder was constituted by five or more of the General Health Questionnaire (GHQ) answers being positive; it thus reflected any mixture of the following: **stress, anxiety, depression, sleeplessness, or confusion**. Recent updates have not shown either a clear upward or downward trend in the prevalences reported by Shidhaya and Patel (2010; the exception to the rule of no trend is that eating disorder prevalence has risen in India; see India State-Level Disease Burden Initiative Mental Disorders Collaborators 2020).

Two strategic decisions were made by Shidhaya and Patel. First they did cognitive interviews (Patel et al. 2008), which convinced them that using the General Health Questionnaire was adequate for an Indian context.

Secondly they cut across disciplinary boundaries in seeking possible ways to explain Indian women's mental disorders. They questioned whether a mental disorder is a medical condition. Many people are living with their mental problems, and especially when we consider the common mental disorders (CMD) hinted at in the GHQ, there may be no medical intervention nor any diagnosis. People may also tend to hide their problems through desirability bias. The team decided to go ahead to see what patterns would emerge in India for mental disorder by social, economic, and female-specific factors. The standard of living proved to be very important.

In Fig. 1.1, the vertical axis is an odds ratio. 'Odds' are a transformation of the probability scale. Odds are the kind of ratio that betting shops and horseracing bets are based upon. The number (N) of events is divided by the number of non-events, or $p/(1-p)$ if p is a probability. The odds ratio is an adapted version of odds. Odds ratios are standardised for a central reference category—here, a type of woman. Odds and odds ratios always lie in the range from zero to infinite, and are continuous variables.

The odds of a woman having a mental disorder in 2003 in India were 1 to 8 (11% of women had one). The odds ratio measures how probable having a common mental

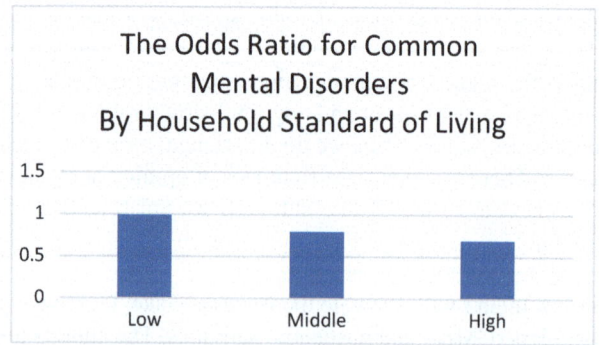

Fig. 1.1 Standard of living and mental disorder among women in India. (Source: Shidhaya and Patel 2010, Table 2. Key: Vertical scale is the self-reported relative odds of having any of the following five disorders at the survey date: stress, anxiety, depression, sleeplessness, or confusion. The baseline odds of 1:1, seen on the vertical scale, are for the low-income households)

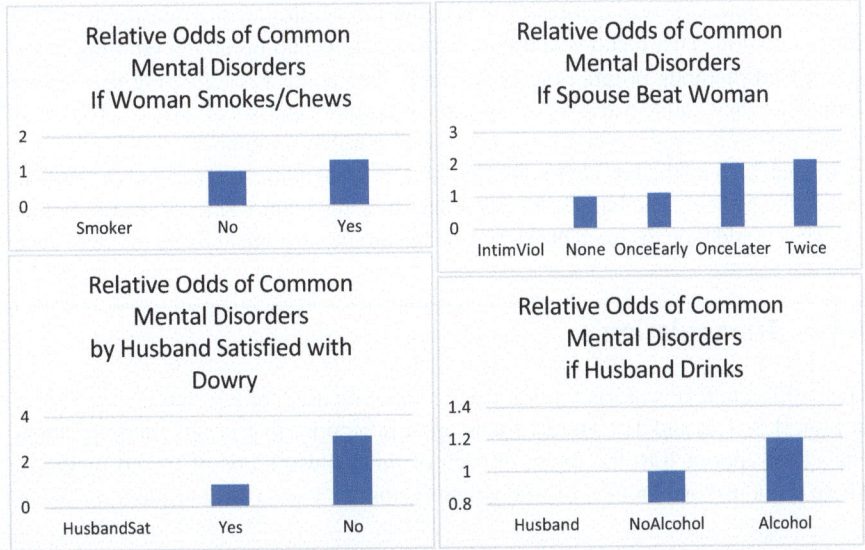

Fig. 1.2 Four factors associated with common mental disorders (CMD) in women in India—a harmonised way to present contrasts. (Source: Shidhaya and Patel 2010, Table 2 for India, 1998–2003. Key: The vertical scale is the odds ratio of having any of the following five self-reported disorders at the survey date: stress, anxiety, depression, sleeplessness, or confusion, relative to the reference group)

disorder (CMD) is, compared with not, given a condition on the horizontal axis. The odds ratios shown are the odds of CMD given different social groups. In Fig. 1.2, Shidhaya and Patel expressed the odds ratio of CMD for other features of society.

Shidhaya and Patel (2010) carefully explained what they meant by modelling these relationships. They argued that 'socio-economic and gender disadvantage factors are independently associated with common mental disorders' (2010: 1510). This was first presented with a mediation diagram which allows both direct and indirect effects from caste group to disorders (see also this book, Chap. 5). But caste group dropped out once the standard of living had been allowed for. A high standard of living was associated with fewer mental disorders in India. The authors therefore omit caste group as an explanatory factor in their interpretations. They also found that gender factors operated 'independently' of other factors. Thus they have reflected upon three facets of the possible models (in India):

– Do women's social gender situations affect their mental state at all? And how?
– Would economic conditions intervene in this process, and if so, how do they intervene?
  Finally,
– What is the best way to portray the effects of social conditions on the prevalence of mental disorders: simple prevalence charts; probability models; or odds models?

They chose the third option which encapsulates the prevalence and the relative probabilities.

Close teamwork was involved in checking the validity of the questionnaire for this population. That result was published separately, and became a building block of a mixed-methods programme of research. The team was reflecting as a group upon what they had learned. What appears as a purely statistical article (2010) was in fact part of a longer term mixed-methods research programme.

In such a case, the bar charts are not merely triangulation. They are an intrinsic part of the project. The bar charts can reflect how the team wanted to test hypotheses. They could help the team build up a clear, complex argument using national data.

1.3   Triangulation

Many different forms of triangulation are commonly used, of which three are shown in Figs. 1.3, 1.4, and 1.5. Here I focus upon reflection, that is self-thinking about one's own approach to the topic, in each of three illustrations of mixed methods. There is not just one method of triangulation, nor just two to three sequences. We do

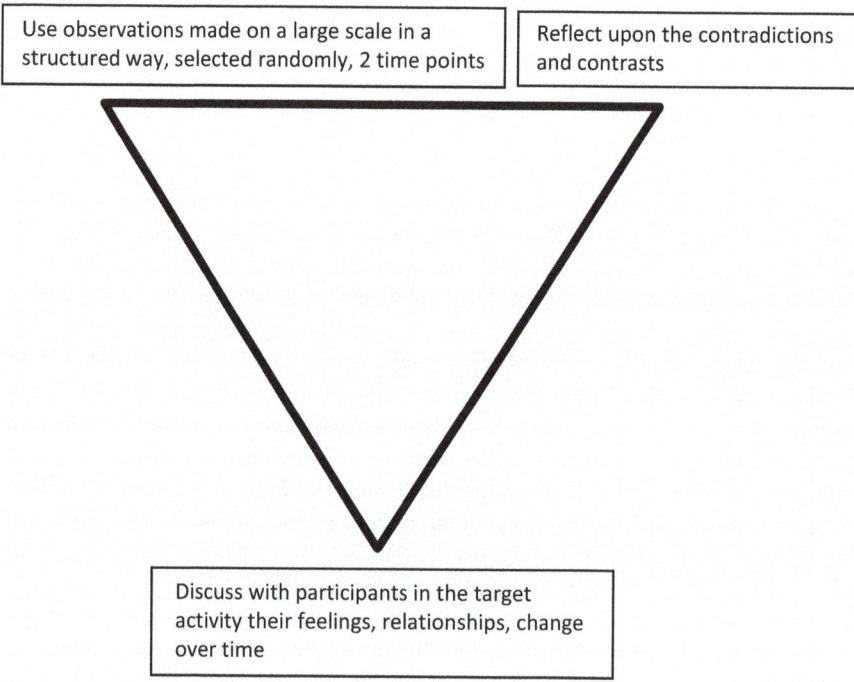

Fig. 1.3 Illustration of triangulation (reflexive SMMR). (Source: The example is based on mixed-methods research which led to Olsen and Zhang (2015); funding was obtained from the British Academy for this analysis. I acknowledge and thank the British Academy for funding Innovation in Global Labour Research Using Deep Linkage and Mixed Methods, grant number PM140147 2014-7, for which I was the principal investigator. I also thank the DFID and ESRC for the grant on Gender Norms, Labour Supply and Poverty Reduction in Comparative Context: Evidence from Rural India and Bangladesh, funded by the Economic and Social Research Council and DFID under grant number ES/L005646/1, 2014-2018)

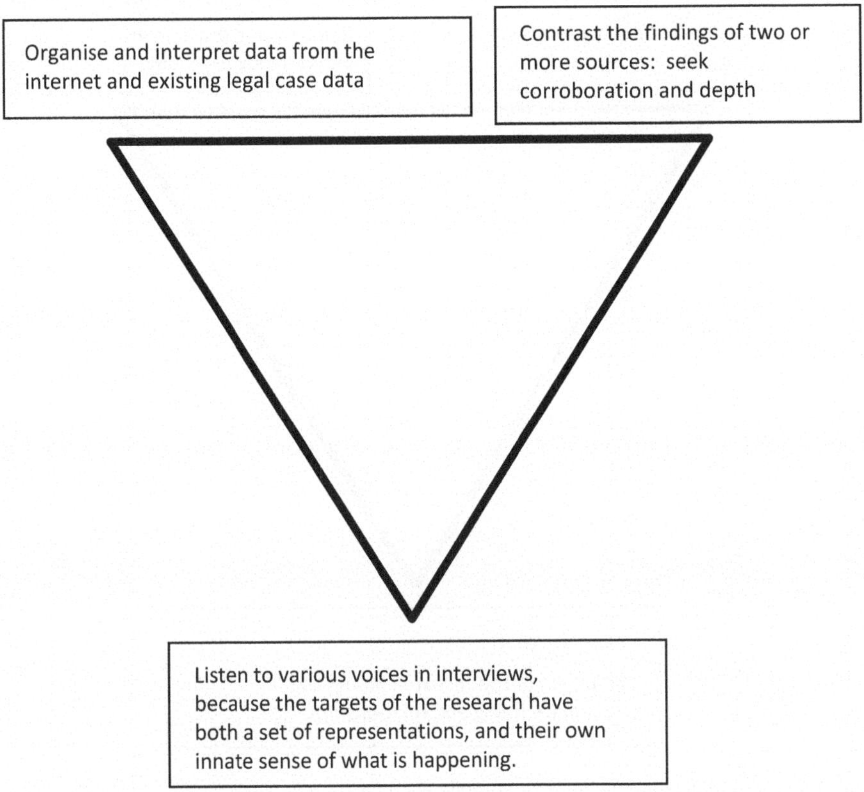

Organise and interpret data from the
internet and existing legal case data

Contrast the findings of two or
more sources: seek
corroboration and depth

Listen to various voices in interviews,
because the targets of the research have
both a set of representations, and their own
innate sense of what is happening.

Fig. 1.4 Illustration of triangulation (SMMR). (Source: Similar lines of thinking appear in Haythornthwaite and Olsen 2018)

a circular, reflective, iterative wheel of possible moves. Each move is aimed at enhancing knowledge.

Figures 1.3 and 1.4 illustrate that the researcher's own standpoint changes during the research. At top right, their own reflexive consideration leads to re-interpretation of both facts and conceptualisations. This is so important that, in this book, triangulation will be taken to always include reflexivity. It is even the case that a project could be entirely desk-based, yet use triangulation; the 'multiple vantage points' that the word 'triangulation' refers to are not necessarily the actual voices of primary fieldwork.

Figure 1.4 illustrates that a legal scholar might use case materials from the courts, combined with interviews. Here there is a mixture of source types, and no quantitative data.

A third form of triangulation is represented in Fig. 1.5. Here triangulation offers opportunities to question evidence that arises freshly in an 'in the field' setting. We visit the field to find out more about the chosen research topic, yet the 'Research Question' is pre-arranged (exceptions being in ethnographic research, and in action research). The research question gives a focus for the time-bound research project (Blaikie and Priest 2013; Philips and Pugh 2010). The research question is open-ended, and the research could lead to surprises. We aim not only at description but also answering a

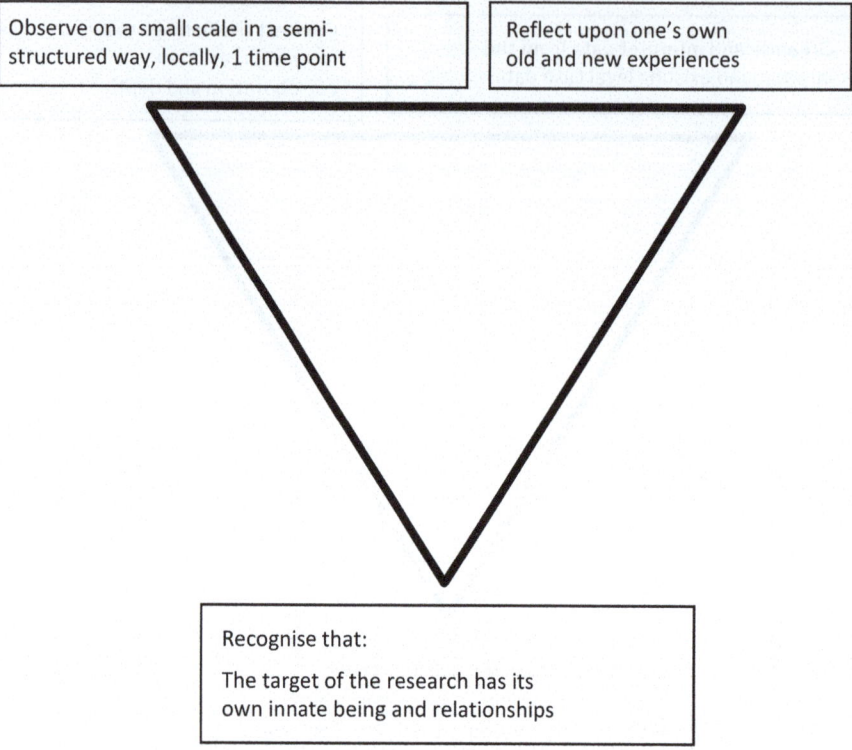

Fig. 1.5 Illustration of triangulation (case-study research)

key question. This answer would make the results original. Triangulation with survey data or national statistics is not necessarily required. On some topics there just isn't any information, for example on how adults feel about children going out to work full-time; or how sex workers perceive the risks of harm during their work. Therefore, one can plan a small foray into a sensitive field to discover more.

A possible problem of the small-scale local fieldwork-based project illustrated in Fig. 1.5 is its small nature. Exceptional conditions can influence the findings while lying tacit and unexpressed. An example is a study of construction workers in Delhi, the capital of India (Khurana 2017). Any study in one locality runs the risk of not being generalisable to other local situations. Such limitations have to be expressed clearly and carefully in the write-up.

As suggested here, reflexivity can be quite personal; it can involve interpreting given data; and it can involve the decision to gather new data or have new experiences.

1.4   Three Domains of Reality, As Realists Approach Research

Next I will turn to how social ontology can enliven a study. After a broad introduction, I focus on distinctions between the real, the empirical and the actual domains.

Open Systems

▶   **Tip** If you apply a depth ontology, there is scope for you as a researcher
to hold views about the topic; to pay attention to multiple stakeholder
voices; and to create arguments about structural aspects of an evolving
system. This often leads to originality: arguing a case which no individual
may have ever 'voiced' before. These achievements also help you link
the quantitative methods with the qualitative parts of a research project.

There are two profoundly opposed positions about ontology. On one side are
proponents of social ontology, who say that to explore the nature of social being is
to do an act that comes prior to all the issues of knowledge and validity that are usually wrapped up in epistemological debates. Here is a sample of the overall approach
of a social ontologist:

> All social material by definition depends on human beings and their interactions and there-
> fore all social stuff, the nature of which is the focus of social ontology, will be of a specific
> geohistorical and cultural form. (Lawson 2016: 962 and see Lawson 2012: 349)

Lawson argues that the location and the date of a study's scope are crucial to
what you find out. The reality, he says, precedes your study and therefore you are
restricted to finding out what is actually present there. My words about this could
seem trite or oversimplified ('actual', 'reality'). I will explain these terms more in
Chap. 4. The framework can be opened out in any concrete arena of research. Thus,
a realist framework is rich and helpful.

One of the key alternative approaches at the constructivist end of the spectrum
involves questioning how points are narrated, how things are worded, and how
descriptions use a select range of concepts (and leave certain crucial things silent).
This polar opposite position to realism would focus on any of the following: narratives, conversations, voices, discourses, tropes, metaphors, analogies, and so on
(Chap. 8). The realist position makes reference to the same things but argues that
these do not exhaust what exists; there are also institutions, structures, agents, dialectics, mechanisms, and causes.

In between these two poles is what I call late-modern social analysis (Chouliaraki
and Fairclough 1999). Within a wide range of disciplines and across the social sciences, people are combining elements of social ontology (what exists) with deeply
well-informed discussions of how people present matters (epistemology). Neither
really 'precedes' the other because they are so different. Any discussion of ontology
must invoke discourses or discourse analysis; any discourse-based study should
consider social ontology.

Society is then an open system. An open system has intrinsic sources of change,
whereas a closed system has mechanical and repetitive workings. An open system
may have one or more of these three forms of openness: (A) open to outside influences, because its boundaries are not very clear, and indeed they are permeable; (B)
open to internal and organic sources of change. One is 'agency'—the capacity for
action, and often it is reflective action with associated cognitions, or discussion—and

if a system has internal agents whose acts can vary then it is likely to be an open system. Any system with reflective agents will be open. Finally, (C) involving complexity. In the realm of complexity we place the overlapping effect of multiple causes, for example; one cause can obstruct the operation of another causal mechanism. Furthermore a causal mechanism might alter *how* another cause works, leading to innumerable minor details of cause and effect within every causal chain, every process, every detailed scene of social activity (Borsboom et al. 2003). Because of the openness of social systems, be careful not to state that they simply and totally reproduce themselves. A class structure does not just reproduce the same class structure. For reasons A, B, and C it produces a new, revised class structure at each point in time.

In Fig. 1.6, which I expand on more in a later chapter, I present some relationships involving explanatory power for structures, structural locations, institutions, memberships, mechanisms, and events. The simple abbreviations used in Fig. 1.6 aim to guide heuristic thinking about a variety of models. The first panel has structure (S), institutions (I), and mechanisms (M) affecting one event. The next two panels allow for change in the structure of institutions, and thus are more ambitious in terms of change over time being explained by the model. The last panel hints at models which reflect change in events (E) such as exam scores, wages, and other outcomes. Following this guide, a simple model of change over time can be

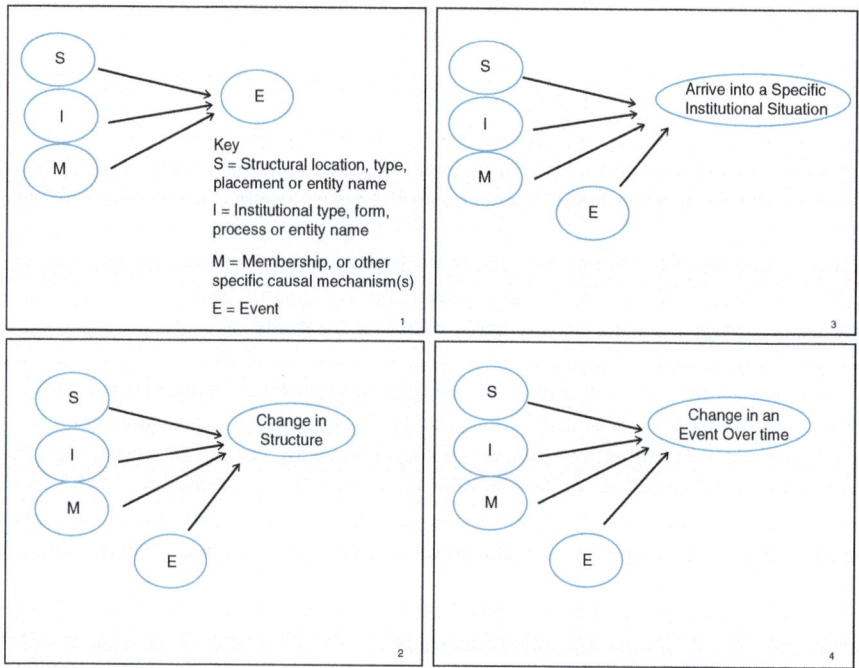

Key: S=structural factors, I=institutional factors, M=mechanisms, including memberships, E=events.

Fig. 1.6 Heuristic diagram of structural-institutional explanatory models: the S I M E model

constructed by putting the latest observed value as the dependent variate and some baseline score as part of the causal model.

Overall, the 'S I M E' model can be a way to think of both fuzzy set analysis and regression models. The S I M E model of explanation marks out an analytical space which the mixed-methods research team is going to augment with concrete findings.

I will explain the realist school of social ontology a little more, because the words used (such as structure) are used in many research projects. We can define structure, institutions, memberships, agency, actions taken, and so on. Before that I will explain the difference between the 'actual' and the reality, and how some authors express the openness of the social system.

Sayer wrote that extensive statistical studies using survey data run a risk of over-generalising and being too abstract (1992, 2000). Yet that has not been my experience. Values do vary over space and time, as Sayer says. For example I found attitudes do vary considerably across the regions of India, but the norms are also coherently clustered within each region. Furthermore, I found on some topics the attitudes of couples are usually homogamous (similar for husband and wife). Attitudes centre around social norms that we can measure on an index scale. A 'region' in India has between 250 m and 550 m people, so attitudes and norms seem to accord with large-scale structures. However, there are broad cultural norms. We can name these; and some people deviate from them. Further structures, such as social class, affect how individual attitudes deviate from local social norms. People's personal experiences and choices also influence their personal attitudes (Olsen and Zhang 2015).

It is an ontological distinction to stipulate that attitudes are not the same as norms. Class is not the same as status and so on. Many scholars successfully stipulate what key terms will mean in their research studies. However, during the research itself they may open up these definitions for scrutiny, revise them, and then explain a refined terminology that better fits the world.

A realist approach to social theory and knowledge can examine structure-agency dynamics. The morphogenetic approach, which focuses on mechanisms of change and of reproduction in a society (Archer 2014), is consistent with the realist use of social statistics. You may have your own approach to how structures, other entities, mechanisms, or people work. There are lots of varieties, and many different schools of thought. There is no need to be a theorist to get going on your project. However, theory and concept development are an important part of the research project, and concepts are not encased in stone from the start.

Distinguishing Reality and the Empirical

One realist approach to science can be described further as follows. Realism would involve a metaphysical stance where a person argues that when people describe things, there is some real world, or universe, to which they are making reference. Therefore obviously the 'subjective' and 'objective' worlds overlap. For example the knowing human is both a subject and potentially an object of knowledge. Secondly, claims about the world have a referential component (what they say about some referred-to parts of the world) while at the same time making oblique reference to the subject's desires, discourses, and interests. This is important for me, because when I study cultural norms and people's attitudes about the same matters, sometimes I have to reflect on my own norms and how they may either cause me to be biased or even blinded, or may help me to understand deviance. As a feminist and a humanist, too, I try to be non-Eurocentric in studying values in various parts of the world. I consider all this normal in the social-science realm in which a statistician or other researcher works.

'Evidence' is not always valid. Validity is not at all simple to establish. Epiphenomenal evidence often gets taken as an adequate representation. Here is an example of an 'epiphenomenal' piece of evidence: a lie during an interview; or a masking statement in a survey. My research shows women in Asia often report 'I am a housewife' and they are economically inactive, but later they report their many activities working with livestock and sewing. Epiphenomena like this are rather common.

I will give one more example of epiphenomena, this time rooted in a poor quality ontology. In water research, it would be naïve to consider an engineer's plans as an adequate account of a whole water supply system. The engineer may have drawn a map and built a model. It might be that engineers need to do planning to conduct water inventions. The interlinkedness of the social world means we manage water whilst also affecting the water cycle itself. Lemon (2003) argued that water systems are often misunderstood by engineers. There is real complexity in the world so the interconnectedness of social and natural systems makes the best action hard to discern in water systems.

Other examples such as explaining divorce, working out what helps children learn to read, and so on would all benefit from having a transdisciplinary theoretical framework. Yet all are necessarily going to have to have imperfect, partial theories in order for theory to be tractable.

The inter- or multidisciplinary ontic starting points are important for social statistics because they provide a basis for statistical work in the interests of strategic, reflexive action. As I set out examples of each of these, the reader can see how helpful the framework is.[2]

---

[2] For some years, discussions in several disciplines have distinguished realism in general from critical realism in particular. Bhaskar would be a typical critical realist, while Sayer would exemplify realism in a broader sense. The differences are discussed in separate works. Among these, notably Archer et al. (1998) explain the transformational aspects of critical realism, while Sayer (1992, 2000) explains some broad enduring principles or metaphysical properties of the world which are highlighted by realists in general. For research projects, Danermark et al. (2002) offer a clearly critical-realist approach to research design, while the position taken by Downward et al. (various dates) is that the critical-realist approach suffuses all realist research. The Downward and Finch idea is that the realist must be a critical realist by virtue of being a critically engaged actor. This view has been opposed by some philosophers on the basis that a category error occurs: the first label, critical realism, is being used to corral a group of scholars, while the second label (realism) is a technical term that focuses upon a wide range of issues of ontology and metaphysics. There is diversity in each school of thought. See Further Reading for sources who discuss this debate.

Three Domains of Reality

Sayer (1992) explains that realism usually claims there are three domains of reality—the 'real', actual, and empirical domains (also Bhaskar 1979, 1993). The real domain has structures whose composition may be a network, hierarchy, or other set of linkages. It is considered separate because not all that is real gets recorded (empirical domain) nor is actualised (potential futures, for example). Secondly the domain of actual events is large, but these are not necessarily observed, nor do we know all about them. Finally at the level of empirics we have data and records, which are the empirical domain. This is a subset or an approximation of some of the actual and real. The empirics quickly become part of the real, but they may be ficts, or in other words, fictional (Olsen and Morgan 2005). Because there are already many debates about these matters, I will make a few comments but leave the matter open to further discussion.

The actual can be forgotten or remembered, but that is not at issue when we ask whether the events or phenomena really 'happened'. People use reflexive memory and dialogues of recall, which may include using theory or applying specific narratives (notably Q&A, autobiography, poetry, specific theories, etc.), to try to recover key aspects of 'the real' when they are going over past actual events. A number of articles on realism help to explain the realist system of thinking which sets out 'actual' as distinct from 'the empirical and the real' (Lewis 2000; Porpora 2015; Olsen 2019). They agree that realism offers a number of linkages of epistemology to social ontology. Realism is thus a form of metaphysics. Other schools of thought may also create linkages, or separations of epistemology from other aspects of theories. I find realism convenient for doing mixed-methods research.

Some realists, like many other social scientists, seek to know about generative causal mechanisms within an open system. The existence of causality in a particular case will need to be demonstrated, not assumed. Thus retroduction may be useful as an investigative method to complement deduction and induction. (See the Glossary and the Further Reading of this chapter.) Retroduction means seeking to know the conditions which have produced the empirical data you have; or more generally, to seek the explanation for how things are. Realists seek among other things to know about the enduring structures but are interested in the institutions, norms, people, marriages, and many other kind of entity. Entity realism is a good entry point for statisticians into realism (Borsboom et al. 2003). But there is much more going on: people's actual experience is too extensive to be described parsimoniously with any degree of accuracy (Quine 1953). Therefore, choice goes on in deciding how to describe a social phenomenon. In setting out these basics, I have tapped into a long debate about whether or not observations can give true information about the world, and whether hypothesis testing could give 'true or false' results (see Further Reading). I tend not to formulate such questions in the previously typical true/false manner. In this book, Chaps. 8 and 9 on warranted arguments help to fill in details of how this is achieved. In Chap. 8 I give examples of deriving findings from multiple, mixed-methods studies. In Chap. 9 I give examples of complex arguments in social science.

Most importantly, realists say, the data are not the same as reality. This difference is what makes operationalisation possible, and it encourages our critical approach to evidence. If we did not think there was any possibility of error, and if the data were all we were studying, then there would be no veil over reality. We might achieve 'proofs'. Most people do think there's a reality, partly hidden, and that data and written records give us partial access to this reality. More importantly, the real has real effects. It can affect what we observe, and thus is profoundly observable. The empirics we observe can be thought of as 'traces' (Lawson 2003). The way reality throws up evidence is called by some a Data Generating Process, or DGP (Hendry and Richard 1990, in Granger, ed., 1990).

The analysis of generative causal mechanisms is widespread; a causal mechanism such as a 'treatment' or 'intervention' can be studied in epidemiology or impact-assessment terms. For a realist the causes are real, whereas to some extent for the DGP approach the causes are unobservable in principle; only traces can be observed. We can also study 'cases' which are ontically distinctive and may exist in nested sets, levels, or as unique ideographic one-off situations (Olsen 2009, Byrne (2009, in Byrne and Ragin, eds., 2009)). In both realist statistics and case-study approaches, the traces are considered good enough to get reasonable knowledge of how causal factors are operating.

The term 'open system' stresses that causes don't work in isolation but may interact with other background factors. Downward and Mearman (2007) discussed the closed-model issue, urging that we use **retroduction** and mixed methods.

Retroduction (See Also Glossary)

Retroduction is moving from what you have in front of you (e.g. empirical data; or data and existing literature) to the reality of what must be true or real in order for the data to appear as they do. It involves moving from data to findings. One does it via some mental machinations. It is a voyage of discovery or realisation, different from deduction (from laws or premises to details) or induction (from detail to general). Retroductive arguments use evidence but are not generally deductive in the way that falsificationist arguments are. See further explorations of retroduction in my earlier book (Olsen 2012) and in a book chapter (2019).

Structural factors are often referred to by realists. A structure is a set of parts, all existing in relation to each other in enduringly patterned ways, such that the whole is an entity that has features beyond the characteristics of the parts. Thus a company has an organisational culture beyond the beliefs or attitudes of the employees. This culture is a social structure. The class system is another structure. It may be exploitative even if individuals or companies do not think they are acting in directly exploitative ways. Structures exist in a holistic, non-reducible way. They have emergent properties beyond their constituent parts, such as 'gentleness' or 'exploitativeness'. A structuralist is someone who attends to the irreducible characteristics, not just the elements which form sub-units of each structure. Structuralists take a holistic view.

Forms of Integration of Mixed Methods

Many authors are recommending integrated mixed methods. Gary Goertz's approach offers a way to integrate the various steps of the research with the ongoing discoveries and firm findings (2017). David Byrne (2002, 2005) argued that we need to avoid the 'variables only' approach commonly found in statistics, and instead, use case-based approaches (also Byrne and Ragin, eds. 2009). Since then a number of authors have advised that we look at small groups of cases, see how they change over time, and name the causal factors even though no statistics will be used. This is known as process tracing. In these studies no 'variables' exist, mainly because they do not aim for an algebraic presentation. However in this book, I will explore how links can be drawn from algebraic models to the 'processes' seen in process tracing. Using variable names or a Boolean algebra offers a practical choice of representation that allows large case-numbers to be handled, offering a contrast with purely qualitative research, and excellent opportunities for triangulation.

When moving to mixed methods a central issue is whether you can remember all the details of all the cases. Using mixed methods we create artefacts of knowledge, such as case summaries, which act as a crutch for the next step of your logical reasoning.

Integrating mixed methods means tracing an argument through various stages of research, similar to how medical researchers do triangulation (O'Cathain et al. 2010). To 'trace' an argument means to use common elements of that argument in different stages, locations, and settings of a research project, and finally to draw conclusions around central parts of the original argument.

1.5   Conclusion

I have argued firstly in favour of distinguishing the real entities from the data. We then have to notice that actual events are often recorded but they are not necessarily the real *underlying* causal mechanisms; we should distinguish mechanisms from contextual factors, use retroduction, and consider systems to be open (permeable, complex, and changing). That is how and why the book advocates a depth ontology.

Analysing the data comes after developing a conceptual framework, introducing some system of measurement, such as categories or operationalisation, and arranging and conducting any fieldwork. Therefore instead of just one big step—such as 'Hypothesis Testing' —I have suggested that the research design can allow for a number of smaller steps. These then lead towards the warranted claims that arise as conclusions of the research.

In the next chapter I will take my logic further: I will set out a variety of logical steps that we use to analyse a situation (moving beyond induction) given this inherent complexity.

Further Reading

This chapter was firstly about triangulation with reflexive thinking; and secondly about basic methodological assumptions of research, which might include realism, agency, and complexity.

Using triangulation is now widespread, even in medical and business research (for medical research, see Bonell et al. 2018; and for management and business, see Molina-Azorín 2011). Mixed methods requires a complex research design and careful sample planning. Teddlie and Tashakkori (eds.) (2003) offer a guide to most of the key steps (also Tashakkori and Teddlie 1998, 2003a, 2003b, and 2009). Creswell and Plano Clark (2018) is a useful textbook. Another textbook at a medium level is Ragin (1987). Ragin explains why using 'variables' is not necessarily the best way to proceed, and how 'cases' can be discerned (see also Ragin 2000, 2008). Creswell and Plano Clark (2007) are widely used by both beginners and advanced researchers using mixed methods (see also Creswell and Plano Clark, eds., 2007). The implications for triangulation for nursing are brought out in a straightforward way by Heale and Forbes (2013).

In the rest of this chapter, looking at basic methodological themes, I have argued for looking at complexity because open systems are real and they bring complexity to most research projects; see Byrne, David (2004) for example on housing and health. Complexity was examined at a general level by Byrne, David (1998), and implications for quantitative methods were drawn out in Byrne (2002) and Bryman (1996). Realist approaches were described by Carter, Bob, and Alison Sealey (2009), Pawson and Tilley (1997), Pawson (2013), and in an explicitly 'critical realist' variant, by Porpora (2015).

For a feminist glance at these issues of realism see Harding (1995) and (1999). She concluded that strategic approaches would be common because of the deep and intrinsic nature of both society and human beings within society.

In discussing realism, the debate about mechanisms as metaphorical tropes is summarised by Williams, M (2019) in a useful edited volume on realism ed. Emmel et al. (eds., 2019). The issue of mechanisms within systems is taken up by Bunge (1997, 2004) and by Elster (1998). In Hedström, P and Swedberg, R (1998a, b) a detailed approach is offered, which might be thought of as realism but not critical realism. One of the key issues here is the multi-layered nature of society and the world, with 'layers' being a metaphor and clearly in danger of limiting our understanding. Another key issue is about the role of the actor, the agent, the researcher, and the research team (agency in general). The realists will tend to see agency as something one may know facts about, while the critical realists may tend to see agency as an essential human characteristic—and like many essentials, something that is not fixed, but is changing and evolving through time. The result of discussions of this topic is that it is now widely agreed that agents who do research are embedded in the system they study. Only under certain very clear circumstances can they consider themselves to be outside the system they are studying (e.g. in experimental conditions in biology).

Acknowledgements I acknowledge and thank the British Academy for funding Innovation in Global Labour Research Using Deep Linkage and Mixed Methods, grant number PM140147 2014-7, for which I was the principal investigator. I also thank the Department for International Development (DFID) and Economic and Social Research Council (ESRC) for the grant on Gender Norms, Labour Supply and Poverty Reduction in Comparative Context: Evidence from Rural India and Bangladesh, funded by the ESRC and DFID under grant number ES/L005646/1, 2014–2018.

Appendix

Table 1.2 A hypothetical example of quantified data on marriage

Id	y	Marriage type	Binary for type 1	Binary for type 2
1	10	1	1	0
2	2	3	0	0
3	13	6	0	0
4	4	1	1	0
5	8	1	1	0
6	6	2	0	1
7	2	6	0	0
Correlation with y:		Low association. Test the mean of y within groups.	0.20	−0.05

The correlations of each type with y are shown in the last row

References

HM Treasury (2020), Government of United Kingdom, *The Majenta Book*, URL https://www.gov.uk/government/publications/the-magenta-book, accessed May 2022.

Archer, Margaret (2014) *Structure, Agency and the Internal Conversation*, Cambridge University Press.

Archer, Margaret, Roy Bhaskar, Andrew Collier, Tony Lawson, and Alan Norrie (eds.) (1998) *Critical Realism: Essential Readings*, London: Routledge.

Bhaskar, Roy (1979) *The Possibility of Naturalism: A Philosophical Critique of the Contemporary Human Sciences*. Brighton: Harvester Press.

Bhaskar, Roy (1993) *Dialectic and the Pulse of Freedom*, London: Verso.

Blaikie, Norman (1993) *Approaches to Social Enquiry*. Cambridge: Polity.

Blaikie, Norman, and Jan Priest (2013) *Designing Social Research. 3rd Edition*, Cambridge: Polity (2nd ed. 2009).

Blaikie, Norman W.H. (2003) *Analyzing Quantitative Data*. London, Thousand Oaks and New Delhi: Sage.

Bonell, Chris, G.J. Melendez-Torres, and Stephen Quilley (2018), "The Potential Role For Sociologists in Designing RCTs and of RCTs in Refining Sociological Theory: A commentary on Deaton and Cartwright", *Social Science and Medicine,* 210, 29–31.

Borsboom, Denny, Gideon J. Mellenbergh, and Jaap van Heerden (2003) The Theoretical Status of Latent Variables, *Psychological Review,* 110:2, 203–219, https://doi.org/10.1037/0033-295X.

Bryman, Alan (1996, original 1988) *Quantity and Quality in Social Research,* London: Routledge.

Bunge, M. (1997) 'Mechanism and Explanation'. *Philosophy of the Social Sciences*. 27 4 410–465.

Bunge, M. (2004) 'Clarifying Some Misunderstandings about Social Systems and their Mechanisms', *Philosophy of the Social Sciences* 34 3 371–381.

Byrne, D. (2005) Complexity, Configuration and Cases, *Theory, Culture and Society* 22(10): 95–111.

Byrne, David (1998) *Complexity Theory and the Social Sciences: An Introduction*, London: Routledge.

Byrne, David (2002) *Interpreting Quantitative Data*. London: Sage.

Byrne, David (2004) "Complex and Contingent Causation: The implications of complex realism for quantitative modelling: the case of housing and health" all in Carter, B and New, C (eds) *Making Realism Work*. London: Routledge 50–66.

Byrne, David, and Ragin, Charles C., eds. (2009) *The Sage Handbook of Case-Based Methods*, London: Sage Publications.

Carter, Bob, and Alison Sealey (2009), "Reflexivity, Realism, and the Process of Casing", Chapter 3 in Byrne and Ragin, eds., *The Sage Handbook of Case-Based Methods*, 2009.

Chouliaraki, Lilie, and Norman Fairclough (1999) *Discourse in Late Modernity: Rethinking Critical Discourse Analysis*. Edinburgh: Edinburgh University Press.

Concato, J. and Horwitz, R. I. (2018) "Randomized trials and evidence in medicine: A commentary on Deaton and Cartwright". *Soc Sci Med*. 210, 32–36. https://doi.org/10.1016/j.socscimed.2018.04.010. Epub 2018 Apr 13. PMID: 29685451.

Creswell, J. W. and V. L. Plano Clark (2007) *Designing and Conducting Mixed Methods Research*, Thousand Oaks, California and London: Sage.

Creswell, John W., and Vicki L. Plano Clark (2018) *Designing and Conducting Mixed Methods Research*, 3rd ed., London: Sage.

Danermark, Berth, Mats Ekstrom, Liselotte Jakobsen, and Jan Ch. Karlsson, (2002; 1st published 1997 in Swedish language) *Explaining Society: Critical Realism in the Social Sciences*, London: Routledge.

Deaton, Angus, and Nancy Cartwright (2018), "Understanding And Misunderstanding Randomized Controlled Trials", *Social Science & Medicine* 210, 2–21, https://doi.org/10.1016/j.socscimed.2017.12.005.

Downward, P. and A. Mearman (2007) "Retroduction as mixed-methods triangulation in economic research: reorienting economics into social science." *Cambridge Journal of Economics*, 31(1): 77–99.

Duvendack, M., Hombrados, J. G., Palmer-Jones, R., & Waddington, H. (2012). "Assessing 'what works' in international development: meta-analysis for sophisticated dummies". *Journal of Development Effectiveness*, 4(3), 456–471. https://doi.org/10.1080/19439342.2012.710642.

Elster, J (1998) 'A Plea for Mechanisms' in Hedström, P and Swedberg, R (eds.) *Social Mechanisms: An Analytical Approach to Social Theory*. 45–73.

Flyvbjerg, Bent (2011) *Making Social Science Matter: Why Social Inquiry Fails and How it Can Succeed Again*, Cambridge, UK: Cambridge University Press.

Funnell, Sue, and Patricia J. Rogers (2011) *Purposeful Program Theory: Effective use of theories of change and logic models*. Sydney: Jossey-Bass.

Goertz, Gary (2017) *Multimethod Research, Causal Mechanisms, and Case-Studies: An Integrated Approach*, Princeton: Princeton University Press.

Goldthorpe, John H. (2001) "Causation, Statistics, and Sociology", *European Sociological Review*, 17:1, 1–20, https://doi.org/10.1093/esr/17.1.1.

Goleman, Daniel (2004), "What Makes a Leader?", *Harvard Business Review*, 82:1, 82–91.

Harding, Sandra (1995) "Can Feminist Thought Make Economics More Objective?", *Feminist Economics* 1, 1: 7–32.

Harding, Sandra (1999) "The Case for Strategic Realism: A Response to Lawson", *Feminist Economics*, 5:3, pp. 127–133.

Heale, R., and D. Forbes (2013) "Research Made Simple: Understanding triangulation in research", *Evidence-Based Nursing*, 16:4, 98–*, https://doi.org/10.1136/eb-2013-101494.

Hedström, P and Swedberg, R (1998a) "Social Mechanisms: an introductory essay" in Hedström, P and Swedberg, R (eds) *Social Mechanisms: An Analytical Approach to Social Theory*. 1–32.

Hedström, P and Swedberg, R (1998b) (eds.) *Social Mechanisms: An Analytical Approach to Social Theory*. Cambridge: Cambridge University Press.

Hendry, D.F., and J. Richard (1990) "On the Formulation of Empirical Models in Dynamic Econometrics", Ch. 14 in Granger, Clive W.J., ed. (1990) *Modelling Economic Series: Readings in Econometric Methodology*, Oxford: Clarendon Press.

India State-Level Disease Burden Initiative Mental Disorders Collaborators (2020), "The Burden of Mental Disorders Across the States of India: the Global Burden of Disease Study 1990–2017", *Lancet Psychiatry*, 7: 148–61, https://doi.org/10.1016/S2215-0366(19)30475-4.

Khurana, Sakshi (2017) "Resisting labour control and optimizing social ties: experiences of women construction workers in Delhi", *Work, Employment and Society*. London, England: SAGE Publications, 31(6), pp. 921–936. https://doi.org/10.1177/0950017016651396.

Lawson, T. (2003) *Reorienting Economics,* London and New York: Routledge.

Lawson, T. (2012) "Ontology and the Study of Social Reality: Emergence, organisation, community, power, social relations, corporations, artefacts and money", *Cambridge Journal of Economics*, 36, 345–385. https://doi.org/10.1093/cje/ber050.

Lawson, T. (2016) "Social all Positioning and the Nature of Money", *Cambridge Journal of Economics*, 40, 961–996. https://doi.org/10.1093/cje/bew006.

Layder, Derek (1993) *New Strategies in Social Research,* Cambridge: Blackwell Publishers. Also Repr. 1995, 1996, Oxford: Polity Press.

Layder, Derek (1998) "The Reality of Social Domains: Implications for Theory and Method", ch. 6 in May and Williams, eds. (1998), *Knowing the Social World*, London: Open Univ. Press, pp. 86–102.

Lemon, Mark (2003) *Exploring Environmental Change Using an Integrative Method,* Hoboken: Taylor and Francis.

Letherby, Gail, John Scott and Malcolm Williams (2012), *Objectivity and Subjectivity*, London: Sage.

Lewis, P. A. (2000). "Realism, Causality and the Problem of Social Structure", *Journal for the Theory of Social Behavior* 30: 249–268.

Masset, E., & Gelli, A. (2013). "Improving community development by linking agriculture, nutrition and education: design of a randomised trial of "home-grown" school feeding in Mali". *Trials, 14*. https://doi.org/10.1186/1745-6215-14-55.

Molina-Azorín, José (2011), "The Use and Added Value of Mixed Methods in Management Research", *Journal of Mixed Methods Research* 5(1) 7–24.

Haythornthwaite, S., and W. Olsen (2018), *Bonded Child Labour in South Asia: Building the Evidence Base for DFID Programming and Policy Engagement*, Dep't for Int'l Development, URL https://www.gov.uk/dfid-research-outputs/bonded-child-labour-in-south-asia-building-the-evidence-base-for-dfid-programming-and-policy-engagement.

Neff, D., and Olsen. W. (2007) "Measuring Subjective Well-Being From A Realist Viewpoint", *Methodological Innovations Online* 2:2.

O'Cathain, A., E. Murphy, and J. Nicholl (2010) "Three Techniques for Integrating Data in Mixed Methods Studies", *British Medical Journal*, 341, https://doi.org/10.1136/bmj.c4587.

Olsen, Wendy (2007) "Pluralist Methodology for Development Economics", *Journal of Economic Methodology,* 14:1, 57–82.

Olsen, Wendy (2009) "Beyond Sociology: Structure, Agency, and Strategy Among Tenants in India", *Asian Journal of Social Science*, 37:3, 366–390.

Olsen, Wendy (2010) "Realist Methodology: A Review", Chapter 1 in Olsen, W.K., ed., *Realist Methodology*, volume 1 of 4-volume set, Benchmarks in Social Research Methods Series. London: Sage. Pages xix–xlvi. URL https://www.escholar.manchester.ac.uk/api/datastream?publicationPid=uk-ac-man-scw:75773&datastreamId=SUPPLEMENTARY-1. PDF, accessed 2021.

Olsen, Wendy (2012) *Data Collection: Key Trends and Methods in Social Research*, London: Sage.

Olsen, Wendy (2019) "Bridging to Action Requires Mixed Methods, Not Only Randomised Control Trials", *European Journal of Development Research*, 31:2, 139–162, https://link.springer.com/article/10.1057/s41287-019-00201-x.

Olsen, Wendy, and Jamie Morgan (2005) "A Critical Epistemology Of Analytical Statistics: Addressing the sceptical realist", *Journal for the Theory of Social Behaviour*, 35:3, 255–284.

Olsen, Wendy, and Min Zhang (2015) *How To Statistically Test Attitudes Over Space: Answers From the Gender Norms Project*, Briefing Paper 3, Gender Norms Project, Creative Commons license, University of Manchester.

Outhwaite, William (1987) *New Philosophies of Social Science: Realism, Hermeneutics and Critical Theory*, London: Macmillan.

Patel V., Araya R., Chowdhary N. et al. (2008) "Detecting Common Mental Disorders in Primary Care in India: A Comparison of Five Screening Questionnaires", *Psychological Medicine* 38, 221–228.

Pawson, R (2013) *The Science of Evaluation: A realist manifesto.* London: Sage.

Pawson, Ray, and Nick Tilley (1997) *Realistic Evaluation.* Sage: London.

Philips, Estelle, and Derek S. Pugh (2010), *How to Get a PhD: A handbook for students and their supervisors,* 5th Ed., London: Open University Press.

Porpora, Douglas V. (2015) *Reconstructing Sociology: The Critical Realist Approach,* Cambridge: Cambridge University Press.

Quine, W. V. O. (1953) *Two Dogmas of Empiricism From a Logical Point of View.* Boston, Massachusetts: Harvard University Press.

Ragin, C. C. (1987) *The Comparative Method: Moving beyond qualitative and quantitative strategies.* Berkeley; Los Angeles; London: University of California Press.

Ragin, C. C. (2000) *Fuzzy-Set Social Science.* Chicago; London: University of Chicago Press.

Ragin, Charles (2008) *Redesigning Social Enquiry: Fuzzy Sets and Beyond,* Chicago: University of Chicago Press.

Sayer, Andrew (1992 (orig. 1984)) *Method in Social Science: A Realist Approach.* London, Routledge.

Sayer, Andrew (2000) *Realism and Social Science.* London: Sage.

Shidhaya, Rahul, and Vikram Patel (2010) "Association of Socio-Economic, Gender and Health Factors With Common Mental Disorders in Women: A Population-Based Study of 5703 Married Rural Women in India", *International Journal of Epidemiology,* 39: 1510–1521.

Smith, Mark (1998) *Social Science in Question,* Sage in association with Open Univ., London.

Tashakkori A and Teddlie C. (1998) *Mixed Methodology. Combining Qualitative and Quantitative Approaches,* Thousand Oaks: Sage.

Tashakkori A., Teddlie C. (2003a) "The Past and Future of Mixed Methods Research: From data triangulation to mixed model designs". In: Tashakkori, A., Teddlie, C. (eds) *Handbook of Mixed Methods in Social and Behavioral Research.,* pp 671–701. Sage, Thousand Oaks, CA.

Tashakkori, A., Teddlie, C., eds. (2003b) *Handbook of Mixed Methods in Social and Behavioral Research,* Sage, Thousand Oaks, CA.

Teddlie C., Tashakkori A. (2003) "Major Issues and Controversies in the Use of Mixed Methods in the Social and Behavioral Sciences". In: Tashakkori, A., Teddlie, C. (eds) *Handbook of Mixed Methods in Social and Behavioral Research.,* pp 3–50. Sage, Thousand Oaks, CA.

Teddlie, Charles, and Abbas Tashakkori (2009) *Foundations of Mixed Methods Research,* London: Sage.

UK Aid Connect (2018) *Guidance Note: Developing a Theory of Change.* Accessed 2019, URL https://assets.publishing.service.gov.uk/media/5964b5dd40f0b60a4000015b/UK-Aid-Connect-Theory-of-Change-Guidance.pdf.

UNDP/Hivos (2011) *Theory of Change. A Thinking and Action Approach to Navigate in the complexity of social change processes.* Author Iñigo R Eguren. For Hivos, The Netherlands, and the UNDP Regional Centre for Latin America and the Caribbean.

Williams, Malcolm (2019) "Making up Mechanisms in Realist Research" in Emmel, N, Greenhalgh, J Manzano, A, Monaghan, M and Dalkin, S (eds) *Doing Realist Research,* London: Sage 25–40.

Mixed Methods with Weakly Structuralist Regression Models

2

Chapter 1 introduced a systematic approach to mixed-methods research and an open-systems logic. My next objective is to show how these diverse components have been put together in a series of excellent mixed-methods research projects (see also Goertz 2017; Beach and Rohlfing 2015). Additionally, in this chapter I show how to use a mixture of logics (Fisher 2004). As a result, each project is unique. I delineate five types of reasoning for empirical projects—induction, deduction, retroduction, logical reasoning, and synthesis. I also illustrate how these fit together. Overall this is a carefully argued, in-depth, and novel approach to research design.

Quite often, the prior work of analysing causes has been done by others, and the researcher arrives at a new project ready to 'do' interviews or 'do' statistics, rooted in this firm foundation of existing knowledge. The topic has been firmly conceptualised, and it is now ready for more work on a new 'operationalisation'. The topic may be ready for explanatory work or a fresh description of change over time. In statistical situations, operationalisation means preparing to measure something, and thus creating either an index or an indicator (Olsen 2012). Measuring something then makes linkages or a mapping from the real to the empirical. Measuring relationships such as correlations is also a form of operationalisation: measuring association. Among case-study researchers an alternative is to create binary indicators or fuzzy sets. The valuable fuzzy-set innovation is treated later on (Chaps. 6 and 7). There is overlap, because any binary indicator can also be thought of as a crisp set: a 0/1 indicator for each 'case' of whether it is 'in' or 'out' of the set described by the detailed variate description. So the statistical school and the qualitative comparative analysis (QCA) schools are not so far apart. Both can also allow for the fact that social change is going on all the time.

It is a good idea to imagine the mixed-methods person as part of a team; part of a lineage of research, perhaps; and building upon what others have done. I would discourage you from taking as 'truth' what others did. It is important to critically assess the social ontology that is being used. Retroduction enables your enquiry to go further than mere social description, to enhance innovation (following the case

W. Olsen, *Systematic Mixed-Methods Research for Social Scientists*,
https://doi.org/10.1007/978-3-030-93148-3_2

study in Morgan and Olsen 2011; the textbook by Danermark et al. 2002; and commentaries by Downward and Mearman 2004, 2007; Downward et al. 2002).

This chapter also shows what strategic structuralism and open retroduction mean in this context.

2.1   Modelling and Methodology for Mixed Methods

▶   **Tip** The variables in a regression do not all have to be of the same 'kind'. They can reflect micro, middle-level, and macro elements as long as each element varies across the sampled cases.

There are a number of mixed-methods exemplars in this chapter. Depending on your background, you may choose one of these to read in detail. The exemplars are presented in a self-explanatory way with simplifications; see below.

Four Systematic Mixed-Methods Research Projects
Cress and Snow (2000)

- Topic: Homelessness Social Movement Organisations (SMOs)
- Place: USA, eight cities,
- Scope: 17 Homelessness organisations
- Methods: case-study approach, institutional ethnography, and qualitative comparative analysis (QCA).

Lam and Ostrom (2010)

- Topic: Watersheds in Nepal
- Place: Nepal
- Scope: 21 watersheds; non-random sample, Engineering and Water and Social Policy discipline areas
- Methods: interviews, historical data, and qualitative comparative analysis (QCA).

Rosigno and Hodson (2004)

- Topic: Fulfilling Work
- Place: English-speaking countries
- Scope: Face-to-face workplaces
- Methods: ethnographic meta-review, systematic comparison, conceptual synthesis.

(continued)

(continued)
   Byrne (Chapter in Byrne and Ragin, eds., 2009)

- Topic: High School Exam Success
- Place: Northeast of England
- Scope: Administrative data, 100% coverage; texts and spreadsheets
- Methods: a study of school inspection reports, with qualitative comparative analysis, and analysis of texts.

Three Exemplars: A Range of Warranted Mixed-Methods Arguments

A **warranted argument** is a phrase associated with Fisher (2001, 2004). Further help with refining arguments is found in Weston (2002) and in Bowell and Kemp (2015). A warranted argument is a more sophisticated thing than a simple theory or a particular claim. Cress and Snow developed a warranted argument using systematic mixed-methods research (SMMR) in their paper on action against homelessness (2000). The exemplars in this chapter mainly used case-based research. First, Cress and Snow (2000) took an innovative research direction by deciding, based on institutional ethnography in eight cities of USA, that there were four outcomes, not just one. My summary diagram (Fig. 2.1) indicates the lines of causality they described in their lengthy, seminal paper (2000).

We can see the causal complexity in multiple ways: in having four outcomes, which would be unusual in a regression-based study; in the definition of the outcomes, which was very refined; and lastly in that the authors allow for mixtures of many different causal variates contributing (jointly) to each outcome, or to its absence. First, Cress and Snow (2000) called the social movement organisations (SMOs) viable or unviable depending on their financial situation. To obtain the viability measure, they examined how the people in the SMOs operated, carrying out face-to-face interviews and shadowing people inside organisations across eight cities. They measured viability as a crisp set (see Chap. 6). They also looked at documents of each SMO that dealt with the homeless people and the issue of their social movements. Altogether they discerned the viability of 17 social movement organisations, though many more exist; this is not a random sample.

Their work on the other causal variates was also innovative. They found that some SMOs dealt with homeless people individually, without developing an agreeing 'framing' to diagnose why homelessness had arisen in their city; others had a clear framing diagnosis. Another dimension of the study was whether each SMO had a prognosis for how to rectify the homelessness problem. Prognosis goes beyond diagnosis, and frames how improvements could be made. The study found that certain SMOs which had both a diagnosis and a prognosis, and were viable, and had two other features as shown in Fig. 2.1, were more likely to succeed in

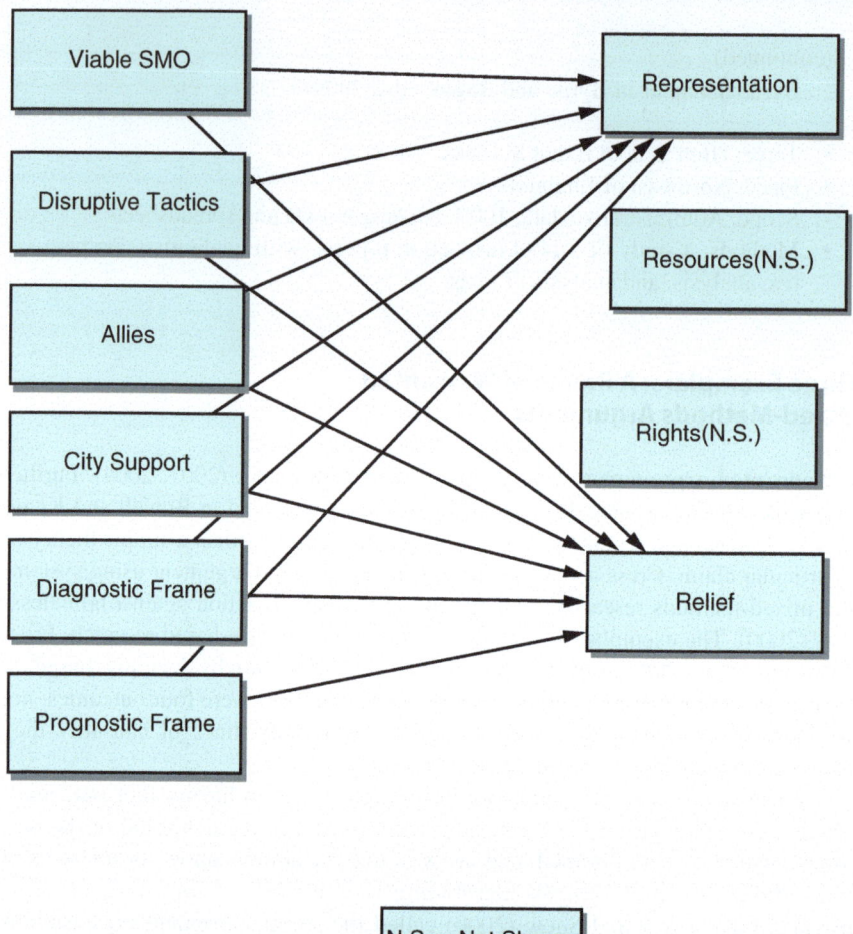

Fig. 2.1 Homelessness social movement organisations' (SMOs) objectives. (Source: Cress and Snow (2000))

improving the representation of homeless people. The point made is that the SMOs of a certain type have a tendency to be capable of improving conditions for homeless people. This typology + tendency approach was noted in Lawson (1997, 2003), who argued that both qualitative and statistical researchers were finding demi-regularities. A demi-regularity, Lawson said, was something we can discover from either small or large samples, and which may however be cross-cut by competing barriers to the outcome: therefore, we do not expect uniformity of the associational relationship. Lawson suggested we can only find 'traces' of causality, using data. The SMO example illustrates finding these traces, partly by using tables, but also generating and then consolidating our in-depth knowledge of the relationships using mixed methods.

Cress and Snow (2000) was a highly pluralist study. Their tables, as seen in Fig. 2.1, were discoveries based on well-focused, yet openly retroductive field research in the eight cities.

This study was unique in offering three alternative versions of the outcome for scrutiny: rights, relief, and resources going to the people who were homeless. They argued the whole debate was not just centred on representation. After doing a summary of the scene, giving factors that explained representation, resources, rights, and relief, Cress and Snow wrote a discussion of each pathway of causality. They argued that most SMOs were unique, but that certain pathways did allow valid comparison. They also noticed similarities.

The concept of a causal pathway is commonly applied in qualitative comparative analysis (QCA; see Chaps. 6 and 7). It overlaps with the notion of process tracing (see George and Bennett 2005, and Bennett and Checkel 2014) as well as concepts in structural equation modelling that enable us to line up several equations, link them together, and form a 'directed acyclic graph' (DAG; see Chap. 4). Directed graphs are also used in sequence analysis and in social network analysis, and all these methods overlap somewhat at the analytical level (visible in Frank and Shafie 2016).

Overall, these various methods are used on a common discursive and ontic basis. They broadly assert that real explanatory mechanisms lie in the society; variables or variates reflect the potential for things to act as causal mechanisms; other entities and conditions lie in the background as conditioning factors that make outcomes contingent; and a series of events can be traced over time through various causal conditions, making 'interaction effects' or pathways of simultaneous or sequential causation on outcomes. In this school of thought, no outcome is a final outcome. An outcome is mainly represented that way for convenience.

Cress and Snow's argument was developed using iteration. Iteration is a mental switch from considering the data collected to considering the findings, then back to the data; and also a teamwork task assignment where the decision is made to send someone back out to the field to make more enquiries, informed by what was found out provisionally. This iterative approach to research design has been urged by Danermark et al. (2002) and Olsen (2012). Further support for iterative stages is found in the grounded theory methods literature (Charmaz 2014). For Snow and Cress (2000), the outcome was expanded to include four outcomes. A similar expansion of the breadth of the study, after the team learned from its experiences in the field, also happened in the case of Lam and Ostrom (2010). Lam and Ostrom wrote an important paper on the watersheds' water supply in Nepal. They had to expand their conception of water supply from short-term supplies to the long-term health of the watershed (see Fig. 2.2). Ostrom won a Nobel Prize for economics for her research on economic norms.

Figure 2.2 also introduces a work which combined over 100 ethnographies into a comparative model. Rosigno and Hodson (2004) carried out a series of integrative studies which synthesised ethnographic evidence about how fulfilling, or unfulfilling, work was for different social groups inside different types of organisations. The team used a variety of sources, all in English, creating a boundary of the

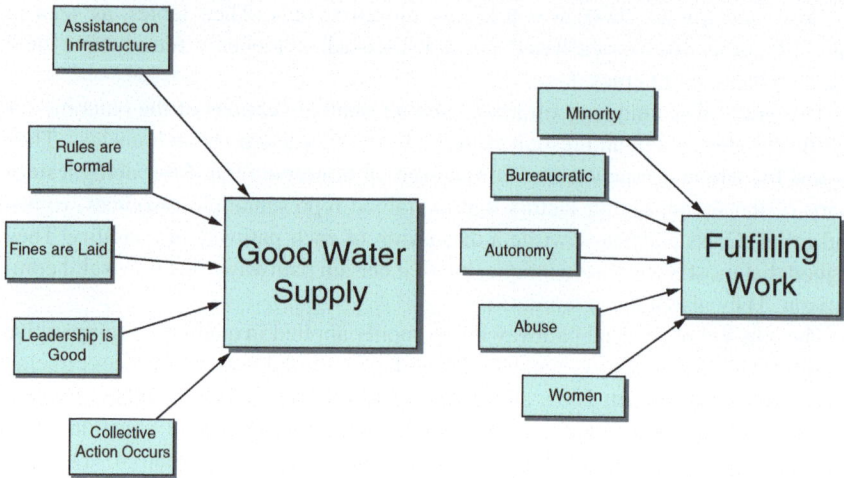

Fig. 2.2 Two exemplars using multidisciplinary qualitative comparative analysis. (Sources: Lam and Ostrom (2010) on the left; and Rosigno and Hodson (2004) on the right)

'English-speaking world' around the scope of the study. Rosigno and Hodson's funding was spread over quite a few years, and gradually a data grid was built up which summarised the ethnographies. This systematised grid of information is rooted in the ethnographic practices of many different scholars.

Both the models shown in Fig. 2.2 highlight structural and institutional factors, as well as events and membership conditions.

During the research project, these authors gradually added extra independent variables to their models. Their argument is that all these factors, represented by variables, work in tandem, or separately, and are necessary and/or sufficient to cause the outcome(s). Discovering whether one mechanism, or one group of conditions which support a different mechanism, was sufficient for the outcome would be valuable. These authors aim to allow for equifinality (multiple pathways). It might be that different pathways also have different 'necessary conditions'. This logic is not so easy to represent in the standard statistical algebra of random variables. Instead, 'necessary' and 'sufficient' causes are reflected in particular patterns in data, dealt with in Chap. 6.

We could call Rosigno and Hodson's method meta-analysis of ethnographies. It was mixed-methods desk research, and it was systematic.

The three exemplars referred to so far are not merely doing induction. Induction is the process of moving from facts about particular cases to a more general statement. It certainly looks like induction when they do ethnography or case-studies. Then we can summarise this with a flow-chart style causal (or explanatory) diagram. I would stress that between these two stages there is a retroductive stage. The authors have asked not only 'what explains Y' but also 'what factors could we introduce that would improve our understanding of how the causal pathways to Y work?' This question is retroductive.

Open retroduction is where we seek wholly new information by returning to the field or getting new data, whereas closed retroduction is where we return to an existing data source for new variates. The open form of retroduction involves re-theorising the whole situation. Bhaskar's concept of meta-critique is one example of retroduction (Bhaskar 1993; Olsen 2010, URL https://www.escholar.manchester. ac.uk/uk-ac-man-scw:75773). Meta-critique refers generally to analysing a theory or discussion from a new or imaginary viewpoint above and surrounding it, that is from a wider point of view. The interaction of meso with micro factors is a key reason why the three examples so far count as meta-critique. Furthermore the interactions of supposedly independent factors, such as 'abuse' and 'women' in Rosigno and Hodson (2004), may not simply be a question of additive or separable additional units of causality. Instead the women's abuse may be part of a causal pathway utterly distinct from men's abuse; in Rosigno and Hodson (2004) there is also a question about the presence of the other gender because none of the production units were wholly female nor wholly male. The complexities usually extend beyond what was recorded in an initial tabular data set. Therefore the open retroduction stage could involve a return to the field, asking for more information by phone, looking up articles, and getting information that we have not originally foreseen that we would need.

Closed retroduction is more familiar. Closed or ordinary retroduction has three options: putting more explanatory or control variates into the causal model; discerning sub-types for the dependent variate which may require separate modelling (sub-group analysis); and introducing various kinds of interaction terms, multilevel terms, or error terms. Thus closed retroduction does not require getting more data, but rather, opening the mind to new possibilities in the form and content of the explanation, using existing data.

In recommending that mixed-methods researchers use open retroduction, I am challenging the idea of the research design being fixed in advance. I will discuss later some implications of having to contemplate changes in the research design, and hence alteration of the research question. If you are a beginner, it could be wise to keep a single project narrow and stick to closed retroduction. For teams, and for large projects and programmes, we can use open retroduction.

Induction sometimes is used as an *ex post* label that shows our conclusions do rest upon evidence in a transparent or sophisticated way. The research process itself is actually more complex.

Inference definitely requires that we make reference to contingent essences. For instance, Rosigno and Hodson and their team were able to point to essences in the causality of fulfilling work which are, in themselves, in turn contingent on underlying social situations. This is not a paradox. Every essence in society is contingent on some features of the social situation that precede it, permeate it, or affect it. These contextual mechanisms are not necessarily deterministic in how they operate. There is always also social change alongside them. There are also natural-world linkages. Thinking of climate change and sustainability issues, for example, which are not prominent in Rosigno and Hodson (2004) but are clearly relevant to Lam and Ostrom's water study (2010), these changes may be invisible or un-felt for a long while. They then can suddenly barge into social consciousness when there is a flood, heat wave, water shortage, or sudden release of pollution. Among complexity theorists, this is known as the causal effect of a unique cause. In general, explanatory analysis is aimed at discerning the diverse causal pathways of some outcome, not reducing any explanation to a single process.

These aspects of multiple causal pathways are covered in *Sociological Methods & Research*, 2018 (47:1) covering the generality, necessity, and sufficiency of causes for given (chosen) outcomes. Papers cover party-system stability and other economic and social topics. These papers cover necessary versus sufficient causality as two distinct things. They have implications for the sampling stage of a project. It often appears that layers or levels in the society need to be considered.

In my own research I once set out three layers for the study of employment (Fig. 2.3). My research on employment illustrates that in studies of work numerous layers all interact with each other. None is dominant; it is also not a strict hierarchy of layers.

The scene is now prepared for making key general points about levels in society. First, nature and society interweave, so neither is 'basic' and neither is superstructure. This results from late modern society's huge impact upon nature, our moulding of tools, walls, and buildings; and our impact upon landscapes. Second the issue arises whether the levels are nested strata, or strata at all, or whether they also interweave. We might break down this issue into key types or examples, such as the social class-gender relation (non-nested); the household-individual relation (nested in Western societies), and the national/local enterprise site relation (nested). Yet examples exist to defy these generalisations. Thus the non-nested relation is the one to which you should be sensitive (Olsen 2009a, b). The non-nested idea is hinted at

Level 3   State Regulation and Welfare Regime (Macro Level)

Level 2   Occupational and Firm-Based Factors (Meso Level)

Operationalised by: Wage Bargaining, Union Roles, Gendered Occupational

Segregation, Labour Market Segmentation, and Firms' Strategies for Using

Cheaper Labour (e.g. Part-Time Workers)

Level 1   Human Capital at the Person-Level (Micro Level)

(individual level) operationalised by: Education Level, Skills, Years of Work

Experience, Years of Service With Current Employer

Fig. 2.3 Three ontic levels in wage determination research. (Based on Olsen 2003, chapter in Downward 2003)

Official Legal Sphere	Enterprise Sphere	Social Sphere
Laws	Implementing Guidelines	Broad culture
Regulations	Enterprise-specific policies	Wider social norms
Implementing Guidelines	Rules (e.g. for lunch break)	Dominant norms
Procedures	Process norms (unwritten)	Contested norms

Fig. 2.4 Non-nested levels in enterprises. Note: The macro was put at the right and the micro in the middle column; but laws and guidelines also constitute macro-level controls, so we have clashes of culture and clashing regulations once we allow for all these informal as well as formal sectoral layers

in Fig. 2.4, showing nesting of enterprises in both the legal and social scene. Figure 2.4 responds to both the Rosigno and Hodson (2004) study and my own studies of workplaces. Yet it is arguable that the social and legal scene also has ways of responding to the enterprise-level events, too. I suggest the levels are non-nested.

In Fig. 2.4, first notice that the enterprise is a rule-taker *vis a vis* official laws. Exceptions exist for those firms operating in the black market or with two sets of accounts. We soon arrive at an interplay over time: more black market leads towards higher probability of detecting some fraud; more fraud leads to new anti-fraud movements; more movements lead towards new laws on fraud; thus the enterprise sphere is not nested 'within' or 'beneath' the legal sphere. Next consider also the social norm situation, which is very complex (shown at the right). For instance 'soft law' is an analysis that recognises localised informal cultural norms. The soft law debate helps legal scholars to analyse the long-term struggles between real social norms and *de jure* laws.

My concept of non-nested cases allows for the mutual conditioning of the N cases (e.g. firms) at one level on the M cases (e.g. regions) on the other level, and vice versa. The classic example of region and industry for firms illustrates this, since a firm can have units in several regions. At the same time, regions embed and link to several firms. So firms do not nest regions, and regions do not simply nest the firms. It is an N * M relationship. The firms' regional location influences their choice of product, and hence industrial focus; but their industrial focus in turn could affect their long-term regional location. We do not find firms entirely clustered in just the perfectly well-suited regions, given their industrial focus, though. Therefore, the influence of regional factors may be diverse, complex, processual, and indeterminate on each firm in each region, and on each firms' industrial focus. Yet the same can be said the other way around: The industrial focus of a firm does not dictate where it is located; many factors related to transport, consumer preference, storage, and the labour force also affect location. We can add more explanatory layers to this N-to-M non-nested relationship if we wish by noting aspects of 'region' in more detail, such as average skill level, the tendency to have in-migrants, coastal status, hilliness, and so on. In my research non-nested relations were common. In statistics we often flatten out this complex structure, placing region, average regional skill level, and other 'variables' alongside the firm-level records to create a firm-level multiple regression analysis, or a series of bivariate tests and tables. Simplification occurs as part of the focusing of the research. The multilevel, deep nested status of the real entities involved is an important lesson to remember, though.

Fourth Exemplar: Qualitative Online Data with Fuzzy Sets, Using Administrative Data

Byrne (2009) analysed a set of all schools in the northeast region of England. He developed an analysis of certain exceptions to the general rule among those schools. First he sets out what he had found to be the general pattern of how (and which) schools have successful grades for their GCSE and O-level high school students. Then he looked at the question of which schools had succeeded, even though they did not have certain key characteristics. Like all schools, these unusual schools had loaded their school reports onto an online database. The project involved download-ing and analysing all these texts, and at this stage Byrne looked selectively at those which appeared unusual compared with the pattern overall. Thus the whole project had these stages: reading up on the literature; gathering the data and finding patterns (induction); describing these and noting exceptions; asking why the exceptions had occurred (retroduction); offering an explanation that encompassed these excep-tional schools (synthesis) (Fig. 2.5).

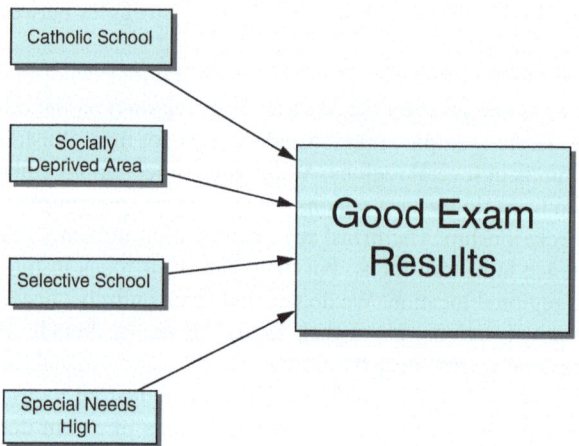

Fig. 2.5 Byrne's analysis of good exam performance in UK schools. Note: The data refer to 126 state secondary schools in the North East Government Office Region of England, based on pub-licly released government data, with some variables binarised using SPSS visual bander (see Appendix Table 2.1). These causal mechanisms do not work in isolation. There is a tendency for deprivation's effect to be inverse (see http://www.ofsted.gov.uk/reports/, accessed Aug. 2020). (Source: Byrne 2009)

The outcome variate in their study was performance across a range of exams among the high school students. The causal factors they considered were numerous (five are shown here).

Using qualitative and quantitative data, this project arrived at a summarising equation in Boolean algebra, then traced those schools which succeeded surprisingly well, and finally offered a finding about these: that student peer review works well to raise average marks of schools. Since peer review had not been in the summarising equation, one might go further and ask that peer review be coded up for all schools. The iteration between project stages has to stop at some point, though. As usual, using Danermark's terms, the feedback loop did stop. As I would describe it, three logics were used.

Induction—retroduction—synthesis took place.

Notice that Byrne et al. took as their multilevel cases the children within schools.

There is no particular requirement to do multilevel modelling purely deductively. As usual we have a series of stages, and we use inductive reasoning when there is independent data or if we have prior results. We can use retroductive reasoning to enliven our interpretations too. We can use mixed methods, and then revisit the statistical models. Here, open retroduction was used to add data to the database covering free school meals and OFSTED school-evaluation texts.

2.2 Strategic Structuralism

One danger with the induction step is a risk of fallacious verification. We tend to verify our ideas by virtue of setting up investigations on the basis of those same ideas. As a result of many doubts and concerns about the fallacy of verification, I promote the idea of a strategic approach to research (also promoted by Harding 1999, under the heading strategic realism). Strategic structuralist research brings a complex set of future possibilities to bear upon the research process, making conscious judgments in a path-dependent process. In particular, a 'strategic' approach is one which considers short-term and long-term effects of the research design decisions, and their costs and benefits (Olsen 2009b). We have to have some clear values and aims, and these help us choose the 'puzzle' we are solving or formulate the Research Question. We then gauge the Research Design from the standpoint of trying to maximise good, and do no harm, while achieving some of these aims. Even if the sampling is poor, and taking into account problems of bias and misrepresentation, we will do the best we can. We cannot entirely avoid the risk of self-verification but we can try to entertain competing ideas.

One can also decide to name and label selected social structures on a strategic basis. Obvious examples are age-group, sex, and ethnic group. Life-cycle stage would be an indicator of a range of institutions, for example child-bearing years or having dependent children: again structural. But which of these will get more emphasis? That's a strategic question.

When I use the term strategic structuralism, I have in mind a contact with many online campaigns which encourage multiple actors to come together (perhaps for a boycott or to choose fair-trade products). I think that the 'actor'-focused approach might ignore structural change. The campaign might fail for lack of a structural strategy. I then try to open up both actor-oriented and structural-change-oriented lines of research.

An example may illustrate strategic decision making further. In the case of taxi drivers, one may divide them into three types (a structural division), and work on each type distinctly because they handle revenue differently, and are taxed differently: Uber drivers (driving on an Uber contract), other self-employed drivers, and drivers who are working for daily or hourly wages. Further structural divisions are possible, for example ethnic, gender, and locality breakdowns, yet eventually one has to stop naming all the structural divisions. The choice of the key factors that are structural is a decision that invokes values as well as being based on evidence.

If I were conducting research on taxi drivers' earnings, I would not just talk to the drivers; I would also consult or construct data tables to show structural differences and relationships.

Realism and 'Weak' Structuralism: Key Terminology
A social structure consists of entities which have long-term relationships between themselves and other entities, and these sets of relationships in turn have emergent properties at some level. For example marriage is a social structure, and families have a structure (e.g. joint or nuclear); the ethnic composition of a city is structural; and the gender division of labour is in many places structural.

On the other hand, institutions refer to sets of normed relationships in society, or between social and natural things. This norming or cultural stereotyping occurs through a mixture of intentional behaviour of agents along with the received structural background, discourses, and language being a part of this background. There are two key uses of the term 'institution' in social science: (1) an organisation; (2) a set of norms about how to do something. The reason the two definitions overlap is that most organisations do have a whole set of entrenched institutionalised behaviours, expectations, stereotypes, and hence norms, which comprise the organisational culture as well as regulation and procedural norms. When a sociologist refers to institutions, however, they mean (2) something a lot wider than (1). An example will help clarify this.

This social-scientific language deviates from a narrower literature in which 'institutions' refer to large organisations such as the United Nations. Institutions refer to sets of norms rather than named or branded organisations. The vehicle-driving institutions, for example, include right-side/left-side driving (depending upon the country); giving way; slowing down and driving past a Stop sign; hand gestures; and many other tacitly known and common behaviour patterns. Out of this short list, only right-side/left-side driving and Stop-sign stopping are regulated explicitly through law. Whilst stopping at a Stop

(continued)

(continued)

sign is regulated, often the law is not followed. Instead people drive through the junction slowly. None of the list of institutions is 'an organisation' in this case, but all of the behaviour patterns are institutionalised. When we move to another country, these patterns are observed to be different, reflecting the different institutionalisation of driving there (in UK, for example, we find large roundabouts and left-side driving).

Strong structuralism refers to when structures of law, society, economics, or politics are asserted to deterministically control and shape other outcomes. Weaker structuralist approaches allow for countervailing impacts. Among these are the competing impacts of simultaneous structural pressures, external agency, and various kinds of habitual, reasoned, or intentional action of agents from within the structures. The word 'weak' refers to having lower dominance of structural features in the theorising of a situation.

The key structural factors in the taxi-drivers' employment research include the ownership of cars, the proprietary rights over the software, and the institutions of law surrounding gig work.

I argued in Chap. 1 that a regression equation has Structural, Institutional, Event, and Membership indicators; we could say the same for a QCA causal model. There will naturally be some overlaps of S and I. In the Appendix I provide a detailed list of variables from a study which had a clear 'independent' and 'dependent outcome' delineation (Cress and Snow 2000). The Appendix to Chap. 4 offers a summary of variables from six studies including that one. Some variables are marked structural but are arguably institutional. Further detailed variables could be added to these short lists. When we turn to statistical examples the 'structural' side emerges very clearly. I gave a clear illustration of this in Olsen (2003 in Downward 2002). Here the dependent variable was employment generation by small enterprises in Ghana. Our research team was able to trace the employment generation achievement of a large sample of rural small-scale industries in Ghana using path analysis. Unusually for economics, interviews and field research combined with surveys led us to creating this path analysis. In recent years, systematic mixed-methods research is becoming more common in economics.

A path diagram is a statistical way to acknowledge multiple causal pathways. Barbara Byrne (2011) shows how to do a path analysis.

However the judgments made in setting up these studies are not only about causality, but also about 'what range of factors shall I show, and what will be prominent as the dependent variable outcome?' To choose 'employment generation' as the outcome imposes a scientific judgment on the project.

Strategic structuralism refers to a researcher or an organisation being strategic about how they plan and conduct survey research and statistical analysis about structures and their sub-entities (Borsboom et al. 2003; and Lawson 1997, 2003, explain the usage of 'entities'). I do not assume these structures are fixed or permanent. So I use a variety of types of evidence. I try not to fall into a naïve trap based on the empirical data. I tend to work from a realist starting-point. I am also convinced that in reality, most statisticians are strategic thinkers (Olsen 2010). In this sense they (we) use canny and wise approaches to resolve tricky scientific problems. We often aim for a long-term value in our solutions, not just a short-term achievement of surface validity or having evidence in the short term supporting our conclusions. There are quite a few structuralists in the statistical sciences: statistics and computing; government statistics; social statistics; geostatistics; medical and biosocial statistics; and psychology. For some, structuralism is similar to demography or the basic assumptions about the use of variables. Yet it is more complex and philosophical than that. It involves giving up on determinism whilst at the same time recognising the holistic 'effect' of things that are much bigger than the micro units in survey data. Furthermore a good structuralist statistician will be looking to discern how the structures are changing in the longer term, and thus is ambitious and has a wide scope for their research programme. Structuralists are very attached to using words like 'structure' but weak structuralists are convinced that (a) structural effects are not permanent nor are they unchanging; and (b) the evidence about the effects of structures' features can be masked by the operation and co-existence of other structures and other entities such as barriers or obstacles.

David Byrne (2009) developed several key arguments in his paper. These included new themes on the background of schools' success, aimed at explaining school assessment outcomes; and also innovations in how we should differentiate and explain the concrete case-wise outcomes themselves. See also the online Annex where Byrne's data is presented at Table A6.5 (www.github.com/WendyOlsen/SystematicMixedMethods).

These were not just debating points: they were warranted arguments. A further discussion of the 'warranted argument' writing style is in Olsen (2019). Warranted arguments are also part of the teachings about 'critical thinking' found in Cottrell (2017) and Bowell and Kemp (2015). Weston (2002) is particularly helpful when starting out with making your own solid arguments. A warranted argument may have to use a 'variable name' to simplify and summarise a causal mechanism. It sounds like a reference to a closed system. But in a warranted argument we can still allow for contextual variation in how things work.

▶ **Tip** You can practice making arguments using illustrations from applied research. It may be helpful to choose one variable from the Appendix to this chapter. The table mentions key variables created by Cress and Snow (2000) from qualitative research and secondary-data enquiries. Try to make notes, developing your ideas about this variable and what Cress and Snow said about it, using full sentences in bullet-point format. Examine this list of sentences to see links between the parts of the argument. To apply the concepts from this chapter, you could annotate the argument with 'inductive step', 'retroductive step' and so on. Mark up any weaknesses in the argument, seen when you find conclusions de-linked from premises, or premises that do not make sense. You can do the same careful study for any key article or book that you want to critique.

There was no reverse causality in Byrne's school study, but there were descriptions of nuanced within-region interactions. Byrne also found that individual factors do not exert cause alone, but rather in tandem with other factors. In the case of Cress and Snow (2000), the factors that were important were not so much 'variables' (things we measure and then represent with algebra) as sets of conditions. These conditions affected each social-movement organisation. Helpful advice on perceiving entities' characteristics in terms of 'set theory' comes from Ragin (1987, 2000; and Ragin 2009, chapter in Byrne and Ragin (2009)). A set is a collection of similar entities. To say a case is in a set, then, is to say it shares key features with other cases in that set. To give a simple example of sets, the types of marriage listed in Chap. 1 form a set: marriage types numbered {1, 2, 3, 4, 5, 6, 99}. In the specialist 'qualitative comparative analysis' literature however, we describe as a 'set' each characteristic of these entities, not just the groupings of the entities themselves. To illustrate, the marriages can be distinguished by the role of the state, the role of a religious leader, the role of a symbolic ring, and so on. These might form three 'sets' for measuring the features of each and every marriage. Thus, each marriage has three or more 'set measurements', indicating the value or type of its features. They can be listed as a matrix:

- Column headings now are marriage features (state, leader, ring).
- Rows consist of marriage details for each case of a particular marriage.
- Row values consist of the set measurement for whether a marriage is 'in' or out or somewhere in between on each feature.
- Types are presented not in the column headings or cell contents, but in patterns that recur.

Many religious marriages for example might appear commonly as having state involved, religious leader involved, and a ring symbolising the marriage. A cohabiting marriage might have just the ring. Set theory and set membership scores offer a rich opportunity for detailed understandings, offering much more than simplistic typologies.

I would like to explain this further now, at a general level.

Byrne and Ragin explain that sets can be either crisp or fuzzy (2009). A crisp set is a simple yes/no for a particular characteristic. That can, in turn, be a set of characteristics. A crisp set data table can be seen as referring to whether each school was in, or out, of the set of those schools having each of a series of particular features. In Chaps. 6 and 7 I explain further about fuzzy membership in sets. If we drew a Venn diagram, the circles would have points 'inside' or 'outside' the circle to reflect crisp sets. Now imagine a grey band instead of a firm circular boundary, which can allow for membership to be 'fuzzy'. Fuzzy memberships are usually measured on a convenient scale from zero to one. Zero means 'no, not in that set' and one means 'fully in that set', but values like 0.3, 0.6 are possible. The role of fuzzy sets was explored by mathematicians long before 'QCA' was invented, and key texts explain some features of set membership patterns. Mendel and Korjani (2018) present a concise synthesis of the pre-extant maths and the current QCA methods; a textbook offering the maths is Bergmann (2008); and a summary is offered in Smithson and Verkuilen (2006).

The exemplars chosen here all used 'set theory' to distinguish which cases were of which kinds. They became well-known studies because they succeeded in doing this in an original way.

A useful book on research design (Blaikie and Priest 2013) helps you set up your project on multiple levels all at the same time, or in sequence (see also Greene 2007).

People can make rather different kinds of claims when arguing a case. For example, some claims present facts, evidence, or data; others express value judgements; others state definitions, criteria, or principles; others give causal explanations; and yet others recommend that we should take a certain course of action. Clearly claims which are of such different kinds have to be evaluated in different ways, so it is important for you to be able to recognise what sort of claim is being made.

2.3   Logics Used in Strategic Structuralist Research

I will briefly set out some of the special logic types that are used in setting up research claims. These logics are used during the research. There is nothing hidden about using all of them. In setting out research conclusions, you may decide to highlight some and not all of these. There is a 'how' stage of research, and then the conclusions reflect 'what' you found. In the 'how' stage, that is how we do research, various kinds of logic are used (see below).

Logics in Strategic Structuralist Research
- **Inductive move**: We gather information and case-studies, or incident descriptions, until enough sampling has occurred to start describing the situation; describe the unit-level patterns; describe the depth ontology in so far as needed to make sense of the key units; conclude.
- **Deductive move**: We theorise, based on existing knowledge; from the theory set out one concretely grounded hypothesis H for a specific place/time; gather data to test this; ensure there is a falsification possibility with these data; decide whether H is falsified or not; conclude.
- **Retroductive move**: We organise or re-state key points of knowledge that are taken as given by a place/time-bound group or are present in the new data; consider why the data look that way and have those patterns; extend this thinking to what are the conditions of possibility of these patterns, and these data; articulate what must or could have caused these patterns to be observed; conclude.

(continued)

(continued)

- **Logical linking**: Based on a series of premises P which are explicit, and some data or evidence E, which are available and can be shared, we move to a conclusion in a syllogism form such that C depends on premises P and evidence E. If necessary, elucidate assumptions A which are at work. Conclude, making sure that C has no elements that arose as hidden premises. Thus, if any of P, E, or A were not the case, then C would not be valid.
- **Synthesising move**: Gather up strands of knowledge claims, argument, data, and conclusions from small- or medium-sized portions of a study; draw it all together; synthesise (gather myriad into a more coherent whole); make a one-page diagram if possible (Miles and Huberman 1984, 1994).

Key: H=Hypothesis, P=Premise, E=Evidence including data, C=Conclusion, A=Assumptions.

The moves are I=Induction, D=Deduction, R=Retroduction, S=Synthesising, L=Logical Linking.

Note: If we are experienced with induction, deduction, and retroduction, there is less need to use synthesis or logical linkage per se. Each is potentially reduced to single steps of logic that have much in common with either I, D, or R. However, after carrying out retroduction, our mind or our data seems to have itself has suggested a finding (logical linkage) or perhaps a new finding has emerged in new wording (synthesis). Examples are found in later chapters. To break down your thinking into these steps is known as parsing the analysis. See Bowell and Kemp (2015) or Cottrell (2013).

Thinking of Snow and Cress's (2000) study, it appears now that the 'journal article format' was a limiting factor. This article was allowed 10,000 words but many journals set an article length limit at 7000–8000 words. The length limit has historically stopped the authors from giving all the details of all the stages of the research.

Snow and Cress conducted institutional ethnography with qualitative comparative analysis. During their work they used all the above logical forms to move towards their conclusions. This book's chapters will tease out other aspects of how people use the synthesising, retroductive, and other logics.

Most readers want to know whether they are testing hypotheses. I have argued earlier that you can use a variety of small logic chunks on your data, aiming to get findings that answer your research question. You may use hypothesis tests, but I advise they not be considered the only way you seek knowledge. The wording of each hypothesis is dependent on many other factors—both social and personal—which affect how we frame our research. You could say, in brief, that the cultural embeddedness of the author of the hypothesis is going to tend to set up the frame of reference; and that language, training, and the academic discipline or setting of the

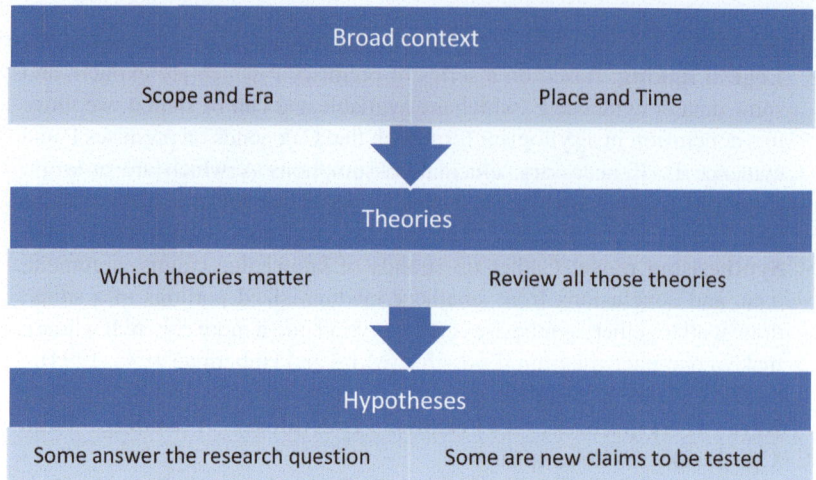

Fig. 2.6 Relevance of theories and hypotheses to mixed-methods research

research are further going to refine and narrow the research. As a result, hypotheses are often quite narrow.

Having a narrow, tight hypothesis is useful because theories are much bigger chunks of explanatory and descriptive claims. You may wish to frame your study broadly as follows (Fig. 2.6).

We can examine case material to see how this issue is worked up in a famous study. In Chap. 4 for example, I look closely at the team projects run by Bourdieu on elite behaviour in France (1979). He looked first at the ways French people sought 'distinction', status, and respect in the society, partly though going out to opera houses and restaurants. Later he also looked at practices around ethnic discrimination. He with his teams considered carefully the ethnicity of the people involved, and the evolution of ethnic categories in the society itself. These interact with social class over time. As social structures, he argued, class and ethnicity are undergoing changes, some of which are gradual. In Chap. 4 I give more detail of the kind of systematic analysis that he did. For his research, using the methods the team chose, the issues of circularity or 'endogeneity' were not much of a problem. In statistical regression endogeneity is widely considered a serious problem since the overlapping of various sets of causes in real, natural conjunctions is usually the case.

A useful book on research design (Blaikie and Priest 2013) helps you set up your project on multiple levels all at the same time, or in sequence. To break down your thinking into these steps of multiple logics, even at multiple levels, is called 'parsing' the analysis. See Bowell and Kemp (2015) and Cottrell (2017).

People can make rather different kinds of claims when arguing a case. For example, some claims present facts, evidence or data, others express value judgements, others state definitions, criteria or principles, others give casual explanations, and yet others recommend that we should take a certain course of action. Clearly claims which are of such different kinds have to be evaluated in different ways, so it is important for you to be able to recognise what sort of claim is being made.

Thus when setting out any hypothesis, you will want to be reflexive about how language and your assumptions will influence it. A classic error in mixed-methods journal articles is to neglect to mention the language and dialect of data collection. If there was a translation stage, the reader needs to know about it. Using Spanish or Bhojpuri or other languages, including English, has an influence in shaping the research. Language itself also has structures and institutionalised norms, and is changing, so we find in the original language the details of metaphor, the tweaking of meaning, and creative description. These are then changed when data or findings are translated into another language.

2.4   Conclusion

So far I have argued that we need to allow for multiple nested types of entities, and processes that link them; we can allow for explanatory and descriptive theories; and we can test hypotheses. In general however the achievement of true findings does not depend only on hypothesis testing. Instead of seeking 'truth' we can seek to make important, valued, warranted claims. The strategic part is moving beyond a narrow, **first-order strategic** realm to take in wider considerations (Olsen 2009b). The **second-order strategies** take into account what others want and what others are saying. The **third-order strategies** go further. We consider how we might improve the situation by influencing what others think and say about that situation. Thus good methodology is not just a question of getting a paper past the peer review stage (first-order strategic thinking). It is not even just about accurate interpretations or good theories responding well to competitors (second-order strategic thinking). It is about placing the research team in the scene and deciding how to pose the interpretation to maximum good effect (third-order strategic thinking; *ibid.*).

Deep linkage refers to linking the data of both qualitative and quantitative types through arguments, so that when interpreting the data the researchers do any or all of the following three triangulating things:

- corroborate arguments built upon different parts of the data;
- complement and build further upon these or other arguments;
- contradict existing theories.

The ambitious researcher also tries to set up new problems to solve, offering insight across the different types of data. Thus, theory triangulation and methods triangulation are presupposed in this three-fold distinction, as proposed also by O'Cathain, Murphy, and Nicholl (2010, 147–148).

Review Questions

Explain what systematic mixed-methods research (SMMR) is, aiming your paragraph at a novice audience.

What would be the 'reflexive' stage in a triangulated study using administrative data, interviews, and statistical modelling?

Explain what is unique about structuralism, giving an example of a social structure.

Explain how retroduction is different from induction.

What is 'open retroduction', and how could you apply it? Explain two ways in which it differs from closed retroduction where you make recourse to the existing dataset.

Further Reading

In this chapter, solidly developed arguments play a key role. The idea has a long track record, and in particular **warranted argument** is a phrase associated with Fisher (2001, 2004). Further help with refining arguments is found in Weston (2002) and in Bowell and Kemp (2015). In the latter, we find distinctions between inductive and deductive arguments, shown both in text and in diagrams, yet I found some difficulties with these (mainly because of begging the question). To avoid begging the question is hard, since we all make methodological and metaphysical assumptions or assertions. Cottrell (2017) is relatively superficial and simplified, compared with philosophical treatments of logic (see e.g. 'Constructivism' or 'Emergent Properties' in the Stanford Encyclopedia of Philosophy, URL https://plato.stanford.edu/contents.html, accessed 2020). Cottrell's book is useful as it contains exercises, ticklists, and key definitions.

A brief summary of mixed-methods possibilities is offered by O'Cathain, Murphy, and Nicholl (2010). This chapter has noted that 'qualitative comparative analysis' will be important in the book, and QCA is summarised by technical experts Mendel and Korjani (2012), and for non-mathematical social science experts by Beach and Rohlfing (2015). The latter also covers 'process tracing' where you study each single case. When learning fuzzy sets and QCA you may choose a book chapter, such as Rihoux and de Meur (2009) found in Rihoux and Ragin, eds. (2009). The two full-length books by Ragin (2008) and Rihoux and Ragin, eds. (2009) both use an easily accessible, non-mathematical mode of presentation.

Appendix

Key: 0 means that a condition is not present. 1 means that a condition is present. These numbers 0 and 1 are also 'crisp sets', that is binary values representing whether that case is in the set of cases which have that particular characteristic. Each case is a social movement organisation (SMO). Four possible dependent variates are shown at the right-hand side.

Table 2.1 Data from Cress and Snow (2000)

Case ID	Causal conditions						Outcomes			
SMO	SMO viability	Disruptive tactics	Sympathetic allies	City support	Diagnostic frame	Prognostic frame	Representation	Resources	Rights	Relief
PUH	1	1	1	1	1	1	1	1	1	1
AOS	1	0	1	1	1	1	1	1	1	1
OUH	1	1	1	0	1	1	1	1	1	1
TUH	1	1	1	0	1	1	0	0	1	1
PUEJ	1	1	0	1	1	1	1	0	1	1
DtUH	1	0	0	0	1	1	1	0	1	0
HCRP	1	1	0	1	1	1	0	0	1	0
BUH	0	1	0	0	0	1	0	0	0	1
DnUH	0	1	1	1	0	0	0	1	0	1
HF	0	0	0	0	0	1	0	0	0	0
HUH	0	0	0	0	0	1	0	0	0	1
HUH	0	1	0	1	0	1	0	0	0	1
MUH	0	1	0	0	0	0	0	0	0	0
HPU	0	0	0	0	0	0	0	0	0	0
MC	0	0	0	0	0	0	0	0	0	0

References

Beach, Derek, and Ingo Rohlfing (2015) "Integrating Cross-case Analyses and Process Tracing in Set-Theoretic Research: Strategies and Parameters of Debate", *Sociological Methods & Research*, 47(1) 3–36, https://doi.org/10.1177/0049124115613780.

Bennett, Andrew, and Jeffrey T. Checkel, eds. (2014) *Process Tracing*, Cambridge: Cambridge University Press.

Bergmann, M. (2008) *An Introduction to Many-Valued and Fuzzy Logic: Semantics, Algebras, and Derivation Systems*, Cambridge, Cambridge University Press.

Bhaskar, Roy (1993) *Dialectic and the Pulse of Freedom*, London: Verso.

Blaikie, Norman, and Jan Priest (2013) *Designing Social Research*. 3rd Edition, Cambridge: Polity (2nd ed. 2009)

Borsboom, Denny, Gideon J. Mellenbergh, and Jaap van Heerden (2003) "The Theoretical Status of Latent Variables", *Psychological Review*, 110:2, 203–219, https://doi.org/10.1037/0033-295X.

Bourdieu, Pierre (1979) *Distinction: A Social Critique of the Judgment of Taste*, Translated from French by Richard Nice, 1984; 1st ed. in Classics Series 2010, London: Routledge.

Bowell, T., & Kemp, G. (2015) *Critical Thinking: A Concise Guide*. 4th ed., London: The University of Chicago Press.

Byrne, Barbara (2011) *Structural Equation Modelling with MPLUS*, London: Routledge.

Byrne, D. (2009) "Using Cluster Analysis, Qualitative Comparative Analysis and NVivo in Relation to the Establishment of Causal Configurations with Pre-existing Large N Datasets: Machining Hermeneutics". Pp. 260–68 in *The Sage Handbook of Case-Based Methods*, D. Byrne and C. Ragin, London: Sage Publications.

Byrne, David, and Ragin, Charles C., eds. (2009) *The Sage Handbook of Case-Based Methods*, London: Sage Publications.

Charmaz, Kathy (2014) *Constructing Grounded Theory*, 2nd ed., London: Sage.

Cottrell, Stella (2013) *The Study Skills Handbook*, 4th Ed., London: Red Globe Press, Macmillan Skills Series.

Cottrell, Stella (2017) *Critical Thinking Skills: Effective Analysis, Argument and Reflection*, 2nd ed., London: Macmillan.

Cress, D., and D. Snow (2000) "The Outcome of Homeless Mobilization: the Influence of Organization, Disruption, Political Mediation, and Framing." *American Journal of Sociology* 105(4): 1063–1104. URL: http://www.jstor.org/stable/3003888

Danermark, Berth, Mats Ekstrom, Liselotte Jakobsen, and Jan Ch. Karlsson, (2002; 1st published 1997 in Swedish language) *Explaining Society: Critical Realism in the Social Sciences*, London: Routledge.

Downward, P., J. Finch, H., and John Ramsay. (2002) "Critical Realism, Empirical Methods and Inference: A critical discussion." *Cambridge Journal of Economics* 26(4): 481.

Downward, P. and A. Mearman (2004) "On tourism and hospitality management research: A critical realist proposal." *Tourism and Hospitality Planning Development* 1(2): 107–122.

Downward, P. and A. Mearman (2007) "Retroduction as mixed-methods triangulation in economic research: reorienting economics into social science." *Camb. J. Econ.* 31(1): 77–99.

Downward, P., J. Finch, H., et al. (2002) "Critical Realism, Empirical Methods and Inference: A critical discussion." *Cambridge Journal of Economics* 26(4): 481.

Fisher, Alec (2001) *Critical Thinking: An Introduction*, Cambridge: Cambridge University Press.

Fisher, Alec (2004) *The Logic of Real Arguments*, Cambridge: Cambridge University Press. (Orig 1988)

Frank, Ove, and Termeh Shafie (2016) "Multivariate Entropy Analysis of Network Data", *Bulletin de Méthodologie Sociologique* 129, 45–63.

George, Alexander L., and Andrew Bennett (2005) *Case Studies and Theory Development in the Social Sciences*, Cambridge, MA: MIT Press.

Goertz, Gary (2017) *Multimethod Research, Causal Mechanisms, and Case-Studies: An Integrated Approach*, Princeton: Princeton University Press.

Greene, Jennifer C. (2007) *Mixed Methods in Social Inquiry*, San Francisco: John Wiley & Sons (via Jossey-Bass Publishers).

Harding, Sandra (1999) "The Case for Strategic Realism: A Response to Lawson", *Feminist Economics*, 5:3, pp. 127–133.

Lam, W. F. and Elinor Ostrom (2010), "Analyzing the Dynamic Complexity of Development Interventions: Lessons from an Irrigation Experiment in Nepal", *Policy Science*, 43:1, pp. 1–25. https://doi.org/10.1007/s11077-009-9082-6.

Lawson, T. (1997) *Economics and Reality*, London: Routledge.

Lawson, T. (2003) *Reorienting Economics,* London and New York: Routledge.

Mendel, Jerry M., and Mohammad M. Korjani (2012) "Charles Ragin's Fuzzy Set Qualitative Comparative Analysis (fsQCA) Used for Linguistic Summarizations", *Information Sciences,* 202, 1–23, https://doi.org/10.1016/j.ins.2012.02.039.

Mendel, Jerry M., and Mohammad M. Korjani (2018), "A New Method for Calibrating the Fuzzy Sets Used in fsQCA", *Information Sciences*, 468, 155–171. https://doi.org/10.1016/j.ins.2018.07.050.

Miles, Matthew B., and A. Michael Huberman (1984, 1994) *Qualitative Data Analysis: An Expanded Sourcebook*, London: Sage.

Morgan, J., and W.K. Olsen (2011) "Aspiration Problems for the Indian Rural Poor: Research on Self-Help Groups and Micro-Finance", *Capital and Class*, 35:2, 189–212, https://doi.org/10.1177/0309816811402646;

O'Cathain, A., E. Murphy, and J. Nicholl (2010) "Three Techniques for Integrating Data in Mixed Methods Studies", *British Medical Journal*, 341, https://doi.org/10.1136/bmj.c4587

Olsen, Wendy (2003) "Triangulation, Time, and the Social Objects of Econometrics", chapter 9 in Downward, Paul, ed. (2003), *Applied Economics and the Critical Realist Critique,* London: Routledge.

Olsen, Wendy (2019a) "Social Statistics Using Strategic Structuralism and Pluralism", in Michiru Nagatsu and Attilia Ruzzene, eds. *Contemporary Philosophy and Social Science: An Interdisciplinary Dialogue*, London: Bloomsbury Publishing.

Olsen, Wendy (2009a) "Non-Nested and Nested Cases in a Socio-Economic Village Study", chapter in D. Byrne and C. Ragin, eds. *Handbook of Case-Centred Research Methods*, London: Sage.

Olsen, Wendy (2009b) "Beyond Sociology: Structure, Agency, and Strategy Among Tenants in India", *Asian Journal of Social Science*, 37:3, 366–390.

Olsen, Wendy (2010) "Realist Methodology: A Review", Chapter 1 In Olsen, W.K., Ed., *Realist Methodology*, volume 1 of 4-volume set, Benchmarks in Social Research Methods Series. London: Sage. Pages xix–xlvi. URL https://www.escholar.manchester.ac.uk/api/datastream?publicationPid=uk-ac-man-scw:75773&datastreamId=SUPPLEMENTARY-1. PDF, Accessed 2021.

Olsen, Wendy (2012) *Data Collection: Key Trends and Methods in Social Research*, London: Sage.

Ragin, C. C. (1987) *The Comparative Method: Moving Beyond Qualitative And Quantitative Strategies.* Berkeley; Los Angeles; London: University of California Press.

Ragin, C. C. (2000) *Fuzzy-Set Social Science.* Chicago; London: University of Chicago Press.

Ragin, Charles (2008) *Redesigning Social Enquiry: Fuzzy Sets and Beyond*, Chicago: University of Chicago Press.

Ragin, C., (2009) "Reflections on Casing and Case-Oriented Research", pp 522–534 in Byrne, D., and C. Ragin, eds., *Handbook of Case-Centred Research Methods*, London: Sage.

Rihoux, B., & Ragin, C. C. (2009) *Configurational Comparative Methods: Qualitative Comparative Analysis (QCA) and Related Techniques* (Series in Applied Social Research Methods). Thousand Oaks and London: Sage.

Rihoux, B., and Gisele de Meur (2009) "Crisp-Set Qualitative Comparative Analysis (csQCA)", in Benoit Rihoux and C. Ragin, eds., *Configurational Comparative Methods: QCA and Related Techniques* (Applied Social Research Methods). Thousand Oaks and London: Sage.

Rosigno, V.J., and Randy Hodson (2004), "The Organizational and Social Foundations of Worker Resistance", *American Sociological Review* 69(1):14–39 https://doi.org/10.1177/000312240406900103

Smithson, M. and J. Verkuilen (2006) *Fuzzy Set Theory: Applications in the social sciences*. Thousand Oaks, London: Sage Publications.

Weston, A. (2002) *A Rulebook For Arguments* (4th ed.). Indianapolis: Hackett Publishing Co, Inc.

Part II

SMMR Approaches in Practical Terms

Causality in Mixed-Methods Projects That Use Regression

3

This chapter begins Part II of the book and thus applies the concepts from Part I to empirical situations. In actual research projects there is usually a research design, a research question, and a set of results. The examples used in Chaps. 3, 4, 5, 6, 7, and 8 are chosen to illustrate a wide variety of social-science research types. This chapter surveys several research examples, summarising their research design and stating the research question for each project. Then a discussion is entered into about how mixed-methods approaches might improve the methodology. The later chapters cover regression models in social statistics (Chap. 4), latent variables and factor analysis (Chap. 5), and qualitative comparative analysis (Chaps. 6 and 7), closing with some primary data-based empirical findings in Chap. 8.

The current chapter offers a systematic mixed-methods approach to regression. Regression models play a key part in many mixed-methods projects. Even a bar chart of means is a very simple 'regression' in the sense that the categorical X variable is implied to influence the mean level of the Y for each grouping or type in X. A S I M E (structural-institutional-mechanisms-events) heuristic framework for regression makes it is easier to consider issues of causality. This chapter's first section presents the S I M E framework in more detail using a multi-layered realist methodology. The second section presents a multilevel causality issue. I show how a 'solution' found in the multilevel algebra actually raises a lot of questions about bias and endogeneity in flat single-level regressions. I offer some solutions using the S I M E framework. I make it explicit how to avoid being methodological individualist(ic) even when you have a flat, single-level regression. The third section offers a more complex example of regression with two equations. I show how a mixed-methods approach might improve upon this research. The example illustrates that the project soon expands beyond what can fit in a single journal article. Therefore we sometimes present the results piece by piece, with a quantitative 'morsel' and later other larger 'morsels', since journal articles have to be short. In conclusion I discuss the useful concept of 'the conditions of possibility' which helps widen a study's explanatory frame.

© The Author(s), under exclusive license to Springer Nature Switzerland AG 2022 59
W. Olsen, *Systematic Mixed-Methods Research for Social Scientists*,
https://doi.org/10.1007/978-3-030-93148-3_3

3.1   Causality in a Regression Model

▶ **Tip** The claim that an X variable is causal on a dependent variable Y is short-hand for saying that a mechanism, or some mechanisms, represented by the X variable have had an effect upon the things represented by the Y variable.

Causality in a regression model is likely to include background knowledge about causes, variables directly reflecting conditions and mechanisms, and other variables which act as proxies for underlying causes (typically like Eq. 3.1).

$$\text{Outcome}_i = f\left(\text{structural, institutional, mechanisms, events,}\dots\right) + e_i \quad (3.1)$$

This equation would be read as: the outcome for each case i depends on the structural, institutional, and other factors including mechanisms and events that apply to each case, or to larger social entities of which it is a constituent part, with an error term e_i. Depending on the discipline, your reading of this S I M E model may vary; for example you could use (S I M) with M for mechanisms in a school teaching situation, or M could present memberships which are a particular kind of mechanism. Further, S E_1 E_2 E_3 could represent your thought that the events part should be blown up into a series of event indicators, such as events many years earlier (such as school absence levels), events in recent years (such as lifelong learning), and recent events (such as night-school attendance). I broadly mean 'events' in a regression are measures that directly indicate a past occurrence of something. An event indicator represents that a particular well-defined event happened. A structural indicator would represent that something complex and enduring existed, such as which neighbourhood the case lives in or whether their area is socially deprived. Institutional measures (such as what type of college the nearest college is: public, private, or mixed) cover socially normal sets of practices. We use a summary code, or a set of codes such as 1, 2, and 3. In regression, a 3-way typology usually becomes a pair of binary indicators. If private and mixed are the two zero-one indicators, then public sector is the base case. The choice of a base case or reference category is important for interpreting the regression and should be set to include plenty of cases rather than a rare type of case.

Variables can reflect a mixture of holistic and directly case-based measures. The case-based measures might be called atomistic, but models should not be atomistic in the broad sense of requiring every variable to be case based. Variables can reflect a sensible mix of proximate (near) causes and distal (distant) causes. The distal causes are more controversial because they may be considered irrelevant, unchangeable, or too 'macro' for regression. Specifically, if a distal cause has uniformity when measured across a whole sample, then one cannot use it in a regression. (It has no variance.) Additionally, a large literature on regression argues that variables in a regression model must not repeat the measurement of some entity. You may find repetition when you insert both proximate and distal causes in one regression, denoted as X_1 (proximate) and X_2 (distal). X_1 may be affected by X_2, so they have similar patterns. An example might be a local regulatory move to promote organic farming, and the farms taking up organic weed control methods. If you repeat the measure, then the two variables X_1 and X_2 are not independent of each other and they will be highly correlated. The result will be a bias in the regression coefficients. Bias is defined as a situation where the slope estimates B_1 for X_1, with and without X_2 in the equation, are different. However there are intermediate levels of correlation where there are good reasons for including two related variables.

In Eq. 3.1, a structural factor is exogenous to a current outcome, meaning that it is not caused by that outcome. An event must in turn be somewhat independent of each structural factor to belong in the equation. For example the person's household social class, sex, marital status, and age group would be measurable structures. Further institutional factors might include indicators of cultural affiliations such as religious background, regularity of worship, or sexual orientation. Specific events relevant to the outcome can also be declared. Thus while the outcome is recorded once per individual, the equation as a whole brings in elements from other types of entities in the transformational structure-agency framework (TMSA).

A S I M E (Structural-Institutional-Mechanisms-Event) Heuristic Framework For Regression

We usually present the regression model as a linear model with a continuous outcome as a dependent variable. To introduce the concept of bias in a slope coefficient, we can consider three models ranging from complex to simple.

$$\text{Model 1 Outcome} = \text{Structural Model} + \text{Agency Model}$$
$$+ \text{Influence Model} + \text{Residual}$$
$$\text{Model 2 Outcome} = \text{Agency Model} + \text{Influence Model} + \text{Residual}$$
$$\text{Model 3}^* \text{ Outcome} = \text{Influence Model} + \text{Residual}$$

*Note: Instead of considering the influence model as a baseline Model 3, one may prefer to posit a simple structural model instead. The risk with this strategy is that one may be seen as a strong structuralist or a structural determinist. Because structural factors are often strong and dominant, however, the data will tend to support you in your simplest Model 3. I have presented an influence model here because we often have unique, well-grounded hypotheses to test as Model 3. Beyond this parsimonious model, we then want to stress the additional gain in explanatory power that Models 1 and 2 offer.

For example if the outcome is 'burnout' among schoolteachers, then the influence model (Model 3) might have two variables for school flexitime policies and whether overtime is paid or unpaid work. For various reasons, schools with either flexitime policies or paid overtime tend to have less burnout. These two variables reflect institutions of the regulation of the labour market in so far as the school has adopted them. In the S I M E approach they fall under the broad heading of institutional factors.

Next look at Model 2, which has agency factors added in. Agency refers to decisions people make, factors which are under one's control, or even the degree of potential (or capacity) people or other actors have to make decisions about things that matter to them. The 'agent' is the actor or any potential actor. Even the school itself could have agency. In the agency part of this model we could imagine two variables: one for the person's uptake of an extra school role, such as head of department or maths leader, and a second variable for the school deciding that all teachers must provide written or typed lesson plans for each lesson. These two variables may be binary in their level of measurement, but that will not matter to my broader point. In the S I M E approach these are going to be considered as 'event' type variables. Memberships also would count here, for example being in the trade union.

The most complex model shown is Model 1 with structural factors added. Here we might have the teacher's gender and age. Another important aspect is their minority ethnic group status and we may also add an interaction effect between some of the pairs of the key variables including minority ethnic group status. With the structural part added, the model may have eight or more variables on the right-hand side. The left-hand side is the dependent variable (see Byrne 2009: 7, 148; Byrne, 1994).

The coefficients are the multiplicative slopes estimated when fitting this model, so that the model best emulates the dataset and minimises the predictions' mean absolute error.

In most social situations the covariances of the structural elements are non-zero. The first structural variable may have a strong impact on the regression. Its insertion will lead to different coefficients in Model 1 compared with the coefficients in Model 3. The second structural variable in Model 1 will also change various other coefficients, compared with Models 2 and 3. The reason for this is that the structural variables also have non-zero covariances with the agentic and institutional variables. The assumption of independence of all the independent variables with each other is usually, in practice, breached.

Some scholars argue for a parsimonious model, perhaps by removing all but one of each type of variable. We could then have one structural (gender), one institutional (overtime), and one agentic (head of department role) variable. Nevertheless, these are still correlated with one another. Men are more commonly holding the role of head of department in schools in many countries; overtime is common among heads of department in schools, and so on.

Even if you choose a small number of coefficients and a small model, there will still be some covariation. It is easy to prove then that the slope coefficient on any single variable will be different in Models 1, 2, and 3. Which one is the best model now becomes an epistemological problem. What values or principles would one use to decide? Besides parsimony other epistemological values would include which is more realistic, which has the best fit, and which model is backed up by other evidence. Other possible criteria may also enter in.

The concept of bias portrays one model as superior and others as biased. When we add too many variables to a model (say 50 or 60 variables which is fairly common), bias is the wrong estimation of one slope due to the co-presence of another variable which is also doing the same job, reflecting the same underlying causal mechanism. To be specific, a model with 35 variables might have $B_9 = 0.56$ and a model with 34 variables might have the same $B_9 = 0.50$. It is not just a case of testing whether these are significantly different estimates. Instead there are fundamental questions here. The usual statistical view is that the larger models are better up to a point, by bringing in new causal mechanisms, but then after that certain optimal point, enlarging the model more will add confusion and cause unwanted, undesirable variations in slope estimates which is why it is called bias.

The S I M E model helps in advising the kind of range of different types of causal mechanism that one might put into a model.

I want to summarise so far about bias. Introducing extra variables that repeat a measurement is not healthy for a regression. But introducing one or more interesting additional variables to reflect freshly discerned underlying causal mechanisms can be very healthy. With large random samples you can add a lot more variables, for example with 10,000 cases you may use 60 variables. Note that some measurements lead to multiple dummy variables. For example: one dummy per region except the reference region, one dummy per ethnic group except the reference ethnic group, and perhaps others, according to the setting.

There is no world in which there is a slope estimate which unambiguously reflects the best model. Here is why. The world of modelling is a small mathematical closed system. Its systemic closure is a figment of our simple algebra. The model is artificial. Modelling is a simplification. Working on such problems, realists have argued that the closure is very artificial and potentially misleading. Doing field research and familiarising yourself with the field of study are crucial to making good interpretations from statistical studies (Downward and Mearman 2007).

My experience with labour-market regression estimates and with studying gender attitudes around work has taught me that structural factors and the intersectionality of structures and agency are broadly important in any large study. Structural factors have proved frequently to be dominant in one sense: if you take them out, the rest of the model changes vastly. Structural factors affect agency, and therefore some agency outcomes (such as joining a trade union), which are inserted as an independent variable, are affected by another 'X' variable. Social, economic, and political variables can be put in a regression together to reflect these diverse effects. You may want to simplify by removing the agentic variables and attributing all the causality to structure. Another option is to remove the structural variables and attribute all the causality to agency. Both these options are wrong, since these two types of entities are related to each other and each should have a great weight in any serious causal argument. A balanced mixture may be the wise choice.

This is notably an ontological argument not just an epistemological argument. It is not about a better fit or improved R-Squared or any other measure of model fit. It is about having the right strategically decided balance of explanatory factors.

The example of schoolteacher burnout is a good one to illustrate how an underlying multilevel model based on a multiply stratified reality could be superior to a simple flat original model. Consider Model 4 which hints at multilevel modelling.

$$\text{Model 4}\left(\text{Level 1} = \text{Persons}, \text{Level 2} = \text{Schools}\right)$$

$$\text{Between Schools}: \text{Burnout} = f_1\left(\begin{array}{c}\text{flexitime policy, average}\\\text{overtime, school size}\end{array}\right)\text{Level 2}$$

$$\text{Within Schools}: +f_2\left(\text{age, sex, overtime, age of head of department,...}\right)\text{Level 1}$$

See also Olsen, 2019. I explain there how a multilevel model can be superior to a one-level model. The language of variation-between and variation-within schools is also used in analysis of variance (ANOVA). The two functions f_1 and f2 are combined in a special multilevel mathematical treatment which has excellent characteristics.

Model 4 can be estimated all at once, along with the correlations of the variables and their multilevel covariance (e.g. larger schools may have older heads of departments). The first level, the School, has common values for its variables across all the cases within the schools. The coefficient on 'age' would be different in Model 4 versus in simpler Models 1, 2, and 3. Therefore one of the models is biased.

To decide which model is superior you can estimate all four and then consider how good the fit is. Once you use a multilevel model, you may get not only a better fit but also very interesting results. The slopes in the first half of Model 4 are school-level responses of burnout to school policies and institutionalised factors. The slopes in the second half of Model 4 relate more to personal factors. These will tend to be smaller once we have the right multilevel model in which some variables are measured at the school level.

The issue of what 'tenets' or norms to use in deciding on a preferred model is broadly a question in the philosophy of science. I argue that with situated knowledge, there are more than two tenets. Not only good fit and good variable choice, but also the strategic choice of what disciplines to challenge and how to integrate disciplines. There are up to ten tenets of good regression (see Chap. 8).

To learn how to create multilevel models, it is best to use a textbook guiding your choice of one particular set of algebraic terms (e.g. either Hox et al. (2017), Snijders and Bosker (2011), or Gelman and Hill (2007)).

An interesting additional learning element arises from the multilevel model. You can obtain the percentage of variation of the outcome at each level. The intraclass correlation coefficient (ICC) offers a way to examine how this percentage changes when we compare two special versions of the multilevel model. As a mixed-methods researcher your main task is to assess whether there is going to be value in moving to multilevel analysis or not.

In the regression stages of research (which occur after the 'Arena Stage'), we have often got to carry out non-perfect estimation of non-deterministic relations using non-perfectly measured variables. A significant issue is—even after allowing for bias and measurement error—that the correlations themselves are disrupted by causal interference from other S I M E elements, either ones which were recorded but left out of this model or ones which were unobserved.

It is best to realise that the interpreters of a model need to use caution. This is harmless good advice. It will prove important in Chaps. 6, 7, 8, and 9. There we have examples of methods which require researcher intervention and researcher interpretation. There is no way to automate the modelling stage or to delegate it to artificial intelligence. Instead, it is a skilled scientific judgement at several key moments.

In this sense, the results are not numeric. The results consist mainly of the claims made using a mixture of words and numbers. The results are hammered out by your team, or crafted by a gifted writer, or carefully honed through 3, 4, or even 10 revisions of a paper segment.

3.2   Stages of Research Design Amendment for Mixed-Methods Research

As a result, when a project has a regression element it really needs not only closed retroduction but also open retroduction. During the stage of 'arena-setting' the scope of research is set. At this stage the variables available might be none, a few or many depending on what secondary data you choose to use. Further secondary data may also be on your desk. The 'arena' stage is when we decide which materials to use, how to plan the research design, and where to operate. See Fig. 3.1 left and middle panels.

Figure 3.1 shows the 'Arena Stage' in the left-hand side, the data stage in the middle, and the creation of conclusive claims as a third stage, at the right.

Without ontological reflection at stage 1, you may make fundamental errors at stage 2. Closed retroduction at stages 2–3 would involve using only variables and texts that already had been gathered. Open retroduction would involve revisiting the scope and data of the study and perhaps going back to the field. By gathering up new data, augmenting what you had, a new model and new interpretations can be created (Olsen 2019).

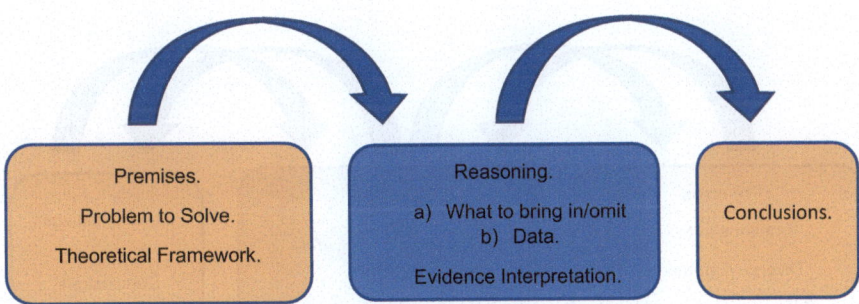

Fig. 3.1 Research stages: the arena stage, data stage, and conclusion

3.3   Deduction Cannot Stand Alone

One implication for philosophy of science is that the use of regression cannot deductively arrive at a sequence of argument-elements leading to verification of a finding. It can only hint at possible truths of potential or likely causality. At the same time, we cannot do regression inductively or in a purely exploratory way. It is crucial to start with a review of literature and not a blank slate, or chaos will ensue. At the same time, it is also very useful to have a feedback loop and a stage of generating new hypotheses. During regression we increase innovativeness by adding variables no-one else has ever thought of inserting. These then generate new findings which require all the other underlying hypotheses to be re-checked (tested), and these new hypotheses to be tested; and we also test the overall model fit.

In a mixed-methods context, the open retroduction might suggest to us going for new survey data. However we often cannot add new variables to an existing survey because the target sample addresses have been lost and/or we do not have ethical permission or informed consent to support revisiting them. Creating a whole new survey is costly, and probably involves a smaller sample size, so this may not be wise. Scholars are currently looking for ways to merge small new data sets with large existing data sets, in case this might help retroduction in the future. (For the moment, you can use entropy-based group weightings if you have a small random set that is similar to, and superior to, a large random set (Watson and Elliot 2015); or you can use propensity score matching if you are testing a treatment effect on a non-random sample or comparative control groups; see Rosenbaum and Rubin 1983.)

To illustrate the open retroduction approach with primary data collection and the single or multilevel model, I have amended the original figure. Now it shows feedback loops. In the methods textbook by Danermark et al. (2002) this is known as iteration. There is great usefulness in returning to the early logics to re-work them, once some results are known. Yet scepticism is also required. Project management is also very important, so that you do not get stuck in a loop or waste a lot of time on repeated bouts of reflection. Use Fig. 3.2 in conjunction with a project management GANTT chart (see URL https://www.pmi.org/pmbok-guide-standards/foundational/pmbok, accessed 2019). Also plan your yearly work objectives with great

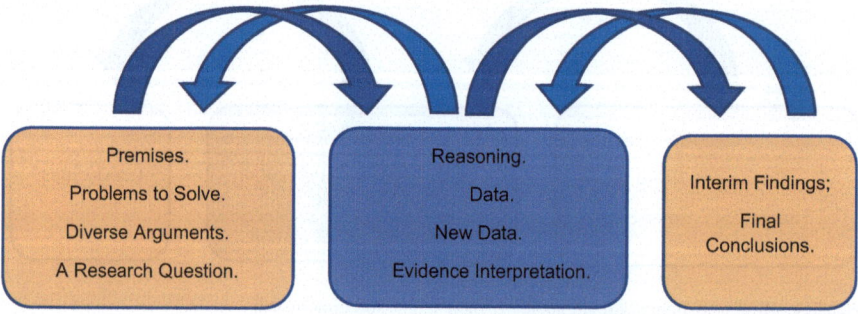

Fig. 3.2 Stages of mixed-methods research with feedback loops of reasoning and discussion

care, so that the feedback-loop approach will not cause you to slow down your career achievement too much. It is meant to improve quality and depth and not to slow the team down.

Figure 3.2 illustrates that retroduction will lead to generating new data, having discussions about the research project's premises, and then creating other forms of feedback loop which might arise from using the mixed-methods data very carefully and then re-working some of the later stages of modelling or interpretation.

It is worth considering what hypotheses actually are, in this context. As a statement-form, they take the form of claims about the world and they can include statements about us-in-the-world. They could be reflexive.

So often, textbooks teach us that the statements we will write as findings are about the world, but do not involve reflexivity. I would urge that we do consider them a product of reflexive thinking and teams having multiple dialogues: findings are social products. Teams produce findings.

It may be easier—or quicker, if not easier—to do open retroduction by using qualitative methods to fill in gaps in knowledge around the new hypotheses. Returning to the regression afterwards, your team will feel confident about how to finalise and publish the models.

We nearly always present these finalised regression models in a reversed order. The largest model is the one that is presented in full first, then smaller sub-models, which are nested, are usually offered. The number of parameters may be large, particularly with multilevel modelling. You could have Model 1 with 65 parameters, Model 2 with 45, Model 3 with 10, and Model 4 with 4. The readers can see the coefficients change as the model is simplified. It also becomes clear how much gain in 'fit' is obtained, relative to complexity, for diverse models.

Some authors in the multilevel modelling school advise an opposite presentational method (Hox et al. 2017). First present an 'empty' model, with the levels declared but with no slope coefficients. Then add the slope coefficient terms gradually, leading towards the largest model. For an experienced reader, these two approaches are seen as mutually consistent since the coefficients are similar, but different, and it is like reading across from left to right for one school of thought and from right to left for the other school of thought.

Overall the widening pool of regression toolkit options is challenging. Details for multilevel modelling can be found in the introductory texts, which are easy to follow but require basic algebra and some past experience with flat regressions (Gelman and Hill 2007). Learning software at an intermediate or expert level (e.g. STATA software) might be helpful for multilevel modelling. To begin with one may like to use Hamilton (2006) or Rabe-Hesketh and Everitt (2009). Gelman and Hill (2007) give a sufficient introduction for most beginners to use the freeware 'R', which operates on a different basis from the typical student software SPSS or STATA. In 'R' software, which is open source and changes regularly through user-contributed command options, the memory holdings of the computer are not restricted to a single table of data. Instead, with 'object-oriented' memory, the computer can hold an immense range of types and sizes of data object at once: tables, numbers, and even diagrams can all be held simultaneously. R's popularity

has accompanied great improvements in computer power. For the easiest introductory software to do multilevel modelling, one might also try the free software
'ML-WIN'.

3.4   A Quantitatively Complex Example

An applied example of the statistical kind is shown next. Bridges, Lawson, and
Begun (2011) explored the relationship of poverty at the household level with the
propensity to do paid work at the personal level. They found that 'for both men and
women extreme household poverty seems to be associated with their personal participation in daily wage employment, whereas in the richer households their participation tends to take the form of self-employment or salaried employment' (2011:
464). Their fuller examination of the nexus of causality around doing work in the
market, including 'paid' self-employment work of men and women, led to the following first regression equation (Eq. 3.2). Here Y* indicates their propensity to do
paid work. It is actually a latent continuous variable, known as a probit, indicating
the probability of the event 'this person works'.

$$Y^*_{i0} = \beta_0 \mathbf{X}_{i0} + \gamma_0 \text{pov}_i + u_{i0} \qquad (3.2)$$

(Bridges et al. 2011: 468)

In Eq. 3.2, bold indicates a vector or list of several terms. Through Eqs. 3.1–3.2,
Bridges et al. offer a labour supply equation for Bangladesh. Here y^*_{i0} is the latent
propensity to participate in the labour market, for case I (a person) in time 0. Pov_i
indicates the extreme-poverty status of the household of person I and pov_i is a structural background factor. It is measured as a yes/no binary (1 = poor, 0 = not poor).
A vector $\mathbf{X}$ comprises a list of other variables. These are other regression variables
that affect labour-market participation (Bridges et al. 2011). Each has a slope $\boldsymbol{\beta}$ and
a corresponding hypothesis that $\boldsymbol{\beta}$ measures how much X affects Y. Age of the person, sex of the head of household, land holding, and local rainfall would be typical
X variables. In terms of the S I M E approach, we have as structural factors the
person's age group, the household's land holding, and local rainfall; an institutional
factor 'gender' reflecting a number of gender-related mechanisms by proxy; and a
dependent variable 'labour market participation', which is an event (Fig. 3.3
illustrates).

In their model, a secondary equation set helps discern factors that affect which
sector a person works in: none, agricultural self-employment, non-agricultural self-
employment, daily waged labour, or salaried employment (ibid.).

$$Y^*_{im} = \beta_m \mathbf{X}_i + \gamma_m \text{pov}_i + u_{im}, m = 0, 1, 2, 3, 4 \qquad (3.3)$$

Here, a case is expected to participate in sector m if y^*_{im} is at its highest level for
i in sector m. The predicted values of Y* are probabilities.

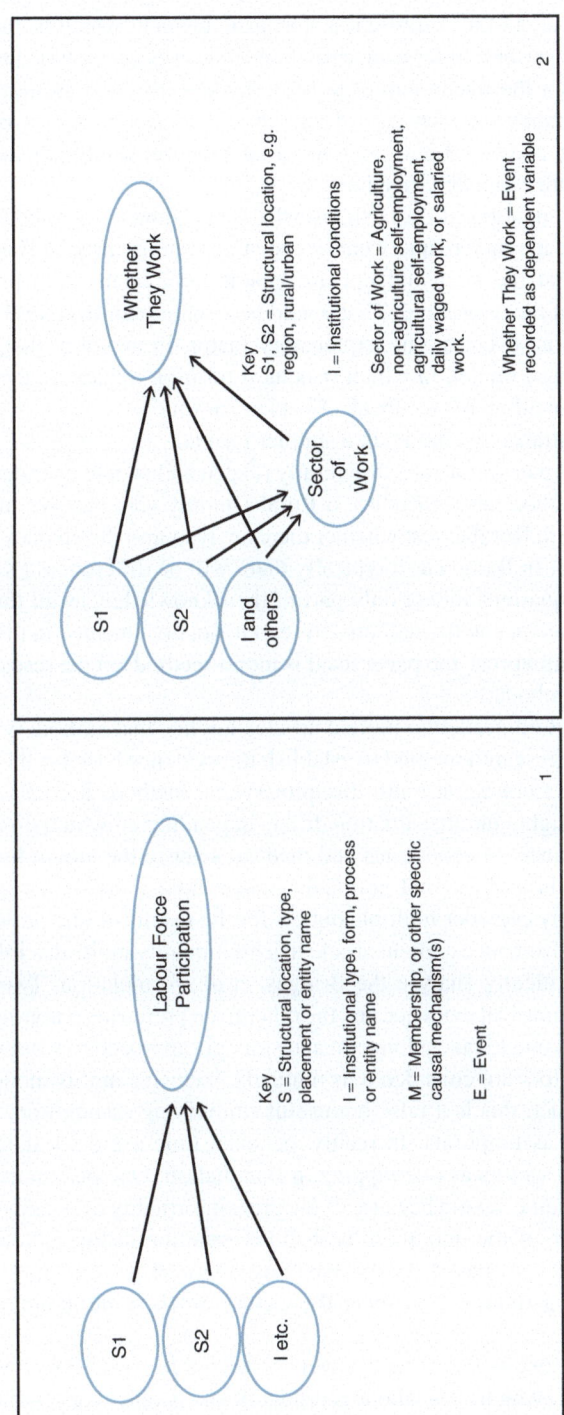

Fig. 3.3 Modelling a mediating factor for a binary labour force outcome

The e_i and the u_{i0} and u_{im} represent a stochastic element which is considered to have a particular distribution for each case. Personal cases are assumed to be homogeneous enough for the linear sum of terms to make sense in presenting us with a prediction or estimate for each case. Equation 3.3 is specified as a multinomial logistic regression and Eq. 3.2 as a latent variable model using generalised linear modelling. They are estimated together.

Although such equations do not exhaust what I can know, or say, about the labour markets, they help me understand what has been happening there in Bangladesh. It was interesting that the team did not use labour force status as a predictor for extreme poverty of the household. That would be a setup with structural position as the outcome, and an event as the explanatory factor. In avoiding that, they used instead a safer causal model, in which structural position influences the event outcome. They did not allow for feedback effects or for time.

Figure 3.3 illustrates this model in a simpler format.

The rural and urban sectors are structurally related and people operate differently in them, so the authors also introduce a Rural Dummy variable. We may interact each X variable with Rural to test whether there are response differences in the rural versus urban areas. In Bangladesh typically there will be differences (Kabeer et al. 2013). Bridges' equations reflect only part of their knowledge, as all three authors have separate experience with qualitative research, not documented in this paper. As happens in many situations, the paper itself is mono-method but the research careers involved mixed methods.

The length limit on academic journal articles implies that authors typically can only use one main research method to establish the evidence base for the main findings of one paper. Looking at multi-disciplinary and methods journals, the length limits can be less tight, and the situation differs depending on which disciplines are involved. For instance, in economics and medical science the length limits can be tight whilst in anthropology and political science more journals accept a longer article. Authors may consider multiple outlets for the results of one project.

Equation 3.3 is from an excellent article in a high-quality multi-disciplinary journal. Yet we can critically analyse the Bridges, et al., formulation. Two criticisms arise—one from empirical evidence and the other from preferring a non-neoclassical theory that enables one to take a feminist and strategic approach to work issues. The first is that the sectors are considered as mutually exclusive in this model. It is not very realistic. In fact, this is a false but useful simplifying assumption. Ideally we would avoid such assumptions. In reality, personal time-use diary data show the work in these four sectors as overlapping in Bangladesh. One person may work in two sectors in the same season because of the great informality of the labour market. As a result, the use of the maximum-probability criterion in Eq. 3.2 can be questioned. It is a criterion based on the labelling approach to principal activity. It ignores subsidiary activities. It assumes the sectors' workers are in mutually exclusive groups.

The second critique is that if theory were restricted to a traditional neoclassical approach to labour (which says labour is drudgery and people work because they're incentivised by human capital and opportunity-cost factors), we would be missing out on the labour obligations that arise from social roles, family norms, and gendered norms about the division of labour. Particularly when some people live in rural areas with partly subsistence economies, the market-based theories of 'human capital' mentioned by the authors have only a partial purchase on the situation.

A third problem is that one can be both a worker and a housewife. I use a wide-ranging theory when I work on labour force participation (e.g. Dubey et al. 2017) so that the labour 'supply' function is redefined as a work-time function. The analysis can be similar to Bridges et al. (2011) but now the underlying theorisation is multi-disciplinary. The work has become economically 'heterodox' in the choice of independent variables.

When we use regression in a project, some of the work can proceed according to textbook advice (De Vaus 2001). Firstly a beginner might run t-tests using survey data to test how influential each social structure is. Secondly at the intermediate level a regression can use the S I M E method (Eq. 3.1) to name the exogenous variables and to place an 'action', that is an event or an agent's response, as a response variable.

The idea of an 'agent' responding to the situation assumes that the 'events' in society are of two main types. Most are caused by a whole range of interacting factors, and thus can be 'explained', but also have a proximate moving force or 'powerful particular' in the person, organisation, or other actor who does an action which constitutes that event. Most events then are carried out by human or non-human agents. An agent is practically defined by being the type of entity that has such a capacity (e.g. a drone can change direction; a group of workers can strike; a husband can walk out). The second main type is much harder to predict or even classify: I mean non-agentic events which occur, and are caused, but for which we don't have an agent who proximately causes them. Examples here are a volcano erupting, a newspaper getting a bad reputation by publishing fake news, or the social equivalent of a flood: the mood among a group of workers worsens noticeably without any specific thing occurring, just a gradual worsening of the organisational culture. I will say more about non-agentic events later (Chap. 10); it would be possible to attribute them to agents but it may be wise to recognise that this may ontically be undesirable. Issues that I will take up at that point are the dialectical nature of the causal capacity of a whole variety of entities, which can lead to non-agentic change; and the role of agents within all this, which can still be powerful, yet not deterministic.

Returning to the possible extensions of Bridges et al.'s work on labour force participation in Bangladesh, we may raise a third issue. In Dhaka in Bangladesh, the supply of clothing for big international companies was produced partly in forced-labour conditions, and the building collapse at Rana Plaza led to the deaths of over 1100 workers. So for more advanced work, invoking discourses that promote change as contrasting with those which preserve the current social scene, the mixed-methods researcher could examine how we could use the causes, motives, and reasons for working—shown in the tables—to find ways to improve working conditions for the garment workers in Dhaka and surrounding cities. We could use qualitative data to explore selected spots, set up focus group in slums, and perhaps engage with stakeholders to develop new action strategies and feedback loops with impactful research. Now the statistical findings may help us to illuminate strategies for change. To support transformative research or action research, new statistical models focused on alternative dependent variables may also be needed. Bangladesh has excellent micro-data at the national level involving surveying of all adults in a random sample. The data, which as in many countries can be obtained without charge, covers both health and lifestyle and labour force topics as well as household structure, migration, education, and household assets.

Therefore, the scenarios for mixed methods with a regression are now three-fold. The project is a smaller timeframe activity with clear planning at the start, for which a 'research design' and a single focused 'research question' may offer guidance, clarity, and narrowness to arrive at a clear finish point in the desired time (Fig. 3.2). Beyond this, the research career will include a range of project- and team-based activities as well as broad scholarly work covering history, language, art, or fiction, as well as non-fiction reading. In the research career, mixed methods and multiple logic are easy to apply but not quite so easy to publish. Compared with the whole career, it may be easier to plan a single research effort using a mixed-methods innovation if you see it as a bounded project.

Beyond these two scenarios, which are both time-bounded, there is also the notion of a school of thought. It is possible to contrast the mixed-methods approaches in a group as one school of thought and to argue that certain schools of thought are founded upon a single core method or methodology. Here we arrive at a large, unmanageable, supra-project entity which we cannot control. Our own agency is crucial to how we try to engage with a school of thought. To make this more concrete I can list five schools of thought, which are related. Suppose we try to discuss scientific work around the concept of school value-added. There are:

- a school of thought focused on measuring student scores before and after special interventions or management-system innovations (Gross Value-Added school of thought);
- a school of thought based on multilevel modelling to discover which level of entity has more explanatory power: school-level holistic factors or child-level factors (a pluralist statistical school of thought);
- a management school based, qualitatively grounded school of thought around school efficacy and efficiency. Efficacy is defined as reaching the goals through strategically planned action and efficiently minimising the costs in achieving the strategic goals;
- a cost-benefit analysis school of thought which attempts to integrate all of the previous three, using mixed logics including induction from qualitative interviews, deduction to test various statistical models, and observational ethnography, shadowing, institutional ethnography or visual ethnography;
- a transformative approach to schooling which sees children as empowered beings which gradually emerge from a dependent status to wholeness and wellness as adults with agency across a wide range of domains of life. In this school of thought qualitative research is central, ethnography or participant observation may play a key role, and school value-added figures are used as secondary data input to the study.

From these five, the first two seem to be firmly grounded in quantitative evidence, the third and fifth seem purely qualitative, but the fourth and fifth have obviously opened up and used triangulation and mixed methods including systematic mixed-methods research.

If it is the case that there are coherent, mutually citing schools of thought that co-exist alongside competing schools of thought, do they accept the validity of each other's findings or not? This question drives us towards an attempt to transcend the limitations of any one particular school of thought.

One writer on schools of thought (Dow 2004) has argued that all the competing theories tap into one reality. Another has pointed out that multiple possible future trajectories create an almost obligatory moral dimension to descriptions of the past and present (Morgan 2013). The concept of 'the conditions of possibility' which retroduction asks us to think about opens up not only the issue of what can explain our present observations but also what present actions can improve the future trajectories. Thus by asking about what conditions could make improvements possible, we open up science's descriptive remit to an additional inherent normative remit.

A quantitative researcher is going to name some variables and measurements, which we may call the modelling part of their research work. The conditions of possibility include the social conditions which make this particular research possible and the conditions which make the variables appear the way they do. Thinking of the model itself, the conditions of possibility include:

- What features of the society have generated this set of institutional and structural features which are influencing the outcome? The answer may involve historical analysis.
- What features of the political and economic situation have generated the specific measurement levels and contrasts which are found in the sample data?
- What were the boundaries of the sample chosen, both regionally and socially? Can these boundaries be relaxed to enable a discussion of macro entities and their history (Lawson 2003)?
- What was the social mechanism of setting the boundaries of this project, and to what extent are generalisations limited within these boundaries due to the sampling that was done?

Examples in social media studies raise issues here about the global versus national internet management schemes. These are a mixture of national government regulations and private-sector country-wise management setups. It may no longer even be possible to study social media postings without a global focus and an internationalised perspective.

Overall then, a researcher is justified in widening the arena of research in order to grasp the situation holistically. Quantitative methods are usually not the only method used, because comparing theories or widening our explanations will tend to lead in a historical direction or suggest to us dynamic tendencies of the scene we are studying.

3.5   Conclusion

In this chapter, I offered a structural-institutional model as a backbone of setting up extended regression models. Working through two examples, I found mixed methods was helpful to understand and interpret the results. After an initial interpretation, a feedback loop can be set up. We see the topic in its historical and dynamic context. The final interpretation may allow for society as an open system. Implicitly my argument is critical of the hypothesis-testing and parsimony arguments in so far as they assume a closed system. I summarised the implications of not assuming 'closure'.

Review Questions

In your own research area, what would constitute a structural-institutional approach to regression?

Taking one of these areas of research (Forestry, Oceanography, Human Resource Management), explain three mechanisms and name a series of events which comprise a sequence but do not have a distinct structure nor a unifying structure.

Is it important to see the world as changing, complex, and open? Explain why or why not, and then name 1–2 social theories or theories about human behaviour and state for each whether it considers the world to be changing, complex, or open?

In statistical models, of which regression is the most common type, a path diagram could be used in which X is a dependent variable, but X then appears as an independent variable in the regression of a different outcome. How is this possible, and why/when does it happen?

Explain in a paragraph what a 'depth ontology' is, giving one example.

The notion of Context+Mechanism→Outcome (CMO) offers a linkage of 'things', which encompass the whole of a social situation, with specific outcomes in that situation. Therefore, do contextual factors belong or not belong in a social survey questionnaire? (Give an example to illustrate your answer.)

Further Reading

This chapter offered a mix of theoretical and practical elements of looking at regression models as explanatory models. Reading further might involve a balanced mixture of these two areas. The main point is that you must carefully word your arguments based on your approach to causality.

On the theoretical side, good starting points are short articles that survey the issues around causality. Goldthorpe (2001) is a balanced, non-technical, non-algebraic summary. Bonell et al. (2018) paper is useful for those readers who are taking a 'treatment approach' or trying to evaluate social interventions. Elster (1998) offers a review of the empirical difficulties for those who are studying mechanisms in a world of mixed, overlapping, multiple mechanisms with self-aware agents who can compensate for known causal mechanisms and thus obviate their appearance in data. Bunge (1997) and Olsen and Morgan (2005) take up the resulting issues for statistics. These include the role of probability in representing causes and the role of confounders. Remember that confounders are important but unmeasured variables which, as indicators of important explanatory factors, if omitted can cause severe problems for the interpretation of regression coefficients. Therefore, we can see how useful these debates are, and a useful treatment for those wanting to use regression is Ullman (2006) because this book chapter covers the manner of adding the confounders to the model. One learns from realising what could be done, even if a particular project does not carry out all these regression tasks using multiple equations (an easily accessible book on using multiple equations is Brown (2015)).

For those wanting practical help, there are both statistical and fuzzy-set summaries that invoke causality. First, in statistics, a good source on causes and explanations is Gelman and Hill (2007). Their book on multilevel modelling is a typical presentation of the idea of caution about causality. They recognise that mechanisms might exist but that correlation cannot prove it, and then show how to explore the data based on clear prior theorising. For more advanced treatment of structural equation models, an exemplar might help (e.g. farmers' preferences and choices were examined in an explanatory, two-step way in Borges et al. (2014)), and a theoretical treatment is found in Deaton and Cartwright (2018, aimed at randomised control trials) or Cross and Cheyne (2018, where interviews about causes are intermingled with demographics using an integrated theory of change). A statistical review is Bollen and Pearl (2013).

For fuzzy-sets, which are taken by the 'QCA' school of thought authors to include issues of necessary and sufficient causality, an online source is found at Ragin (2008, URL http://eprints.ncrm.ac.uk/1849/1/Resdisgning_social_inquiry.pdf (sic) or more extensively annotated materials at CORE.ac.uk, a training site: Ragin, no date, https://core.ac.uk/download/pdf/229986009.pdf in easily accessible presentation slide format). See also Smithson and Verkuilen (2006), which shows how fuzzy-set superset relations can be applied to quantitative data.

Danermark et al. (2002) created the detailed plan of the feedback-loop approach to research, taking a critical realist point of view. Field (2009, 2013) has implicitly taken a broadly realist perspective in his textbook on how to use the popular software package 'SPSS'. Field's books give details of pre-regression hypothesis tests, notably paired t-tests, and he also describes mediation tests in regression which are useful for diagnosing endogeneity. These tests all depend upon closure for both internal and external validity. The closure debate was summarised by Downward et al. (2002). To put it in a simple way, closure involves a model which covers every relevant explanatory factor, and closure is required when we use algebra to avoid the risk that a parsimonious model will have unmeasured, confounding factors. Lawson (2003) critiques the use of algebra in economics partly on the basis of the impossibility of closure in an algebraic model.

Because models never have closure in the strict sense, most statisticians are very cautious about interpreting a regression as a causal explanation. In Goldthorpe (2001), however, the clear distinction is made between interpreting variables as being linked with firm associations (which he calls the robust dependence approach) versus considering the things in society and not just the 'variables'. He calls this latter approach the generative process approach. This distinction is very important (Olsen 2012). If you only look at variables, your ontology is thin; but if you are using variables as traces to examine a deeper or wider society, then your statements are going to be about society and its generative processes, or mechanisms, not about the variables. Thus even in 2001 it was explicitly acknowledged that the thin approach, which appears these days in many data science textbooks and data-driven exploratory methods, was at risk of leading the research teams towards unsupported generalisations. To summarise material from later chapters of this book, the thick approach is known as a depth ontology (Chap. 8) and the thin approach often appears as methodological individualism, but there is much to gain from learning about these matters in detail.

To avoid the closure problem, add to your knowledge the concepts of process tracing (Beach 2018) and designing complex projects (Blaikie, 2000; Blaikie and Priest, 2013). To gain a sense of how science can sometimes progress through an encompassing step, see Cook (1999). The encompassing step Cook describes occurs when a new model takes up the explanatory and descriptive work of an old model but also covers a wider range of situations or offers more and better explanations.

References

Beach, Derek (2018) "Achieving Methodological Alignment When Combining QCA and Process Tracing in Practice", *Sociological Methods & Research*, 47(1) 64-99. https://doi.org/10.1177/0049124117701475.

Blaikie, Norman W.H. (2000) *Designing Social Research: The logic of anticipation*. Cambridge, UK; Malden, MA: Polity Press: Blackwell.

Blaikie, Norman, and Jan Priest (2013) *Designing Social Research*. 3rd Edition, Cambridge: Polity (2nd ed. 2009).

Bollen, Kenneth, and Judea Pearl (2013) "Eight Myths About Causality and Structural Models", pages 301–328 in S.L. Morgan (Ed.), *Handbook of Causal Analysis for Social Research*, London: Springer.

Bonell, Chris, G.J. Melendez-Torres, and Stephen Quilley (2018), "The Potential Role For Sociologists in Designing RCTs and of RCTs in Refining Sociological Theory: A commentary on Deaton and Cartwright", *Social Science and Medicine,* 210, 29-31.

Bridges, Sarah, David Lawson, and Sharifa Begum (2011) "Labour Market Outcomes in Bangladesh: The Role of Poverty and Gender Norms", *European Journal of Development Research*, 23:3, 459-487.

Brown, Timothy A. (2015) 2nd ed. *Confirmatory Factor Analysis for Applied Research*, NY and London: Guilford Press.

Bunge, M. (1997) "Mechanism and Explanation". *Philosophy of the Social Sciences*. 27(4), 410-465.

Byrne, Barbara (1994) "Burnout: Testing for the Validity, Replication, and Invariance of Causal Structure Across Elementary, Intermediate, and Secondary Teachers", *American Educational Research Journal*, 31, 645-673.

Byrne, D. (2009) "Using Cluster Analysis, Qualitative Comparative Analysis and NVivo in Relation to the Establishment of Causal Configurations with Pre-existing Large N Datasets: Machining Hermeneutics". Pp. 260–68 in *The Sage Handbook of Case-Based Methods*, edited by D. Byrne and C. Ragin, London: Sage Publications.

Cook, S. (1999) "Methodological aspects of the encompassing principle." *Journal of Economic Methodology* 6: 61-78.

Cross, Beth, and Helen Cheyne (2018) "Strength-based approaches: a realist evaluation of implementation in maternity services in Scotland", *Journal of Public Health* (2018) 26:4, 425–436, doi.org/10.1007/s10389-017-0882-4.

Danermark, Berth, Mats Ekstrom, Liselotte Jakobsen, and Jan Ch. Karlsson, (2002; 1st published 1997 in Swedish language) *Explaining Society: Critical Realism in the Social Sciences*, London: Routledge.

De Vaus, D. A. (2001) *Research Design in Social Research*. London: Sage.

Deaton, Angus, and Nancy Cartwright (2018), Understanding And Misunderstanding Randomized Controlled Trials, *Social Science & Medicine* 210, 2–21, https://doi.org/10.1016/j.socscimed.2017.12.005.

Dow, S. (2004) Structured Pluralism, *Journal of Economic Methodology*, 11: 3.

Downward, P. and A. Mearman (2007) "Retroduction as mixed-methods triangulation in economic research: reorienting economics into social science." *Cambridge Journal of Economics*, 31(1): 77-99.

Downward, P., John H. Finch, and John Ramsay (2002) "Critical Realism, Empirical Methods and Inference: A critical discussion." *Cambridge Journal of Economics* 26(4): 481.

Dubey, A., W. Olsen and K. Sen (2017), "The Decline in the Labour Force Participation of Rural Women in India: Taking a Long-Run View", *Indian Journal of Labour Economics*, 60:4, 589–612. URL https://link.springer.com/article/10.1007/s41027-017-0085-0

Elster, J (1998) "A Plea for Mechanisms" in Hedström, P and Swedberg, R (eds.) *Social Mechanisms: An Analytical Approach to Social Theory*. 45–73.

Field, A. (2009) *Discovering Statistics Using SPSS for Windows: Advanced techniques for the beginner*, London: Sage.

Field, A. (2013) *Discovering Statistics Using IBM SPSS Statistics*, London: Sage, 4th ed.

Gelman, Andrew, and Jennifer Hill (2007) *Data Analysis Using Regression and Multilevel/ Hierarchical Models*, Analytical Methods for Social Research Series, Cambridge: Cambridge University Press.

Goldthorpe, John H. (2001) "Causation, Statistics, and Sociology", *European Sociological Review*, 17:1, 1–20, https://doi.org/10.1093/esr/17.1.1.

Hamilton, Lawrence C. (2006) *Statistics with Stata*, London: Brooks.

Hox, Joop, Mirjam Moerbeek, and Rens Van De Schoot (2017) *Multilevel Analysis: Techniques and Applications*, Third Edition (Quantitative Methodology Series), London: Routledge.

Kabeer, Naila, Lopita Huq, and Simeen Mahmud (2013) "Diverging stories of "missing women" in South Asia: is son preference weakening in Bangladesh?" *Feminist Economics*, 20:4, 1–26, https://doi.org/10.1080/13545701.2013.857423

Lawson, T. (2003) *Reorienting Economics*, London and New York: Routledge.

Morgan, Jamie (2013) "Forward-looking Contrast Explanation, Illustrated using the Great Moderation", *Cambridge Journal of Economics*, 37:4, 737–58.

Olsen, Wendy (2012) *Data Collection: Key Trends and Methods in Social Research*, London: Sage.

Olsen, Wendy (2019) "Social Statistics Using Strategic Structuralism and Pluralism", in *Contemporary Philosophy and Social Science: An Interdisciplinary Dialogue*, edited by Michiru Nagatsu and Attilia Ruzzene. London: Bloomsbury Publishing.

Olsen, Wendy, and Jamie Morgan (2005) "A Critical Epistemology Of Analytical Statistics: Addressing the sceptical realist", *Journal for the Theory of Social Behaviour*, 35:3, 255–284.

Rabe-Hesketh, Sophia, and Brian Everitt (2009) *A Handbook of Statistical Analyses Using STATA*, London: Chapman and Hall/CRC.

Rosenbaum, Paul R., and Donald B. Rubin (1983), "The Central Role of the Propensity Score in Observational Studies for Causal Effects", *Biometrika*, 70:1, pp. 41–55. URL http://links.jstor.org/sici?sici=0006-3444%28198304%2970%3A1%3C41%3ATCROTP%3E2.0.CO%3B2-Q

Rossi Borges, João Augusto, Alfons G. J. M. Oude Lansink, Claudio Marques Ribeiro, and Vanessa Lutke (2014), "Understanding Farmers' Intention to Adopt Improved Natural Grassland Using the Theory of Planned Behavior", *Livestock Science* 169, 163–174. 10.1016/j.livsci.2014.09.014 1871-1413.

Smithson, M. and J. Verkuilen (2006) *Fuzzy Set Theory: Applications in the social sciences.* Thousand Oaks, London: Sage Publications.

Snijders, Tom A.B., and Roel Bosker (2011) *Multilevel Analysis: An Introduction To Basic And Advanced Multilevel Modeling*, 2nd rev. ed., London: Sage.

Ullman, J. (2006), ch. 17 of Tabachnick, B.G. & Fidell, L. S., eds., *Using Multivariate Statistics* (4th Ed). Boston: Allyn and Bacon. See also Ullman (2006) "Structural Equation Modeling: Reviewing the basics and moving forward", in Statistical Developments And Applications section, *Journal Of Personality Assessment*, 87(1), 35–50, URL https://pubmed.ncbi.nlm.nih.gov/16856785/

Watson, Samantha K., and Mark Elliot (2015), "Entropy balancing: a maximum-entropy reweighting", *Quality and Quantity* 50(4):1–17. https://doi.org/10.1007/s11135-015-0235-8

Multiple Logics in Systematic Mixed-Methods Research

4

Regression modelling can be seen as one step in a larger methodology, systematic mixed-methods research (SMMR). SMMR would refer to any project using a mixture of methods, of which some are qualitative and some are quantitative and hence systematic. By summarising five exemplars where authors have created explanatory models, this chapter shows that it is common to use multiple logics. All these logical moves are common in SMMR and indeed in social research overall. For each project, I first summarise what was done, then examine how a mixed-methods research design might have further developed each project. For example in labour supply research there can be a role for *institutional ethnography*, which is a holistic research approach. This chapter also considers *semi-structured interviews*, *case-comparative methods*, *closed* versus *open retroduction*, and *multiple correspondence analysis (MCA)*. Systematic methods offer useful summarising techniques that bring together correlated variables and supplementary variables, alongside your case studies or interviews.

> ▶ Defining Realism   Realism is a philosophical approach which argues that some entities pre-exist and influence the social researchers when we try to carry out research. Realists also allow for things that are 'macro' entities beyond any one individual's control, and there are many key points in realist literature on methodology (Danermark et al. 2002). In particular, they point to three domains of the world: real, actual, and empirical. The real world is complex, changing, and open to change over time. The actual domain is a sequence of innumerable events and things, including processes over time. However the empirical realm, where records are kept, may represent the real and the actual in

© The Author(s), under exclusive license to Springer Nature Switzerland AG 2022
W. Olsen, *Systematic Mixed-Methods Research for Social Scientists*,
https://doi.org/10.1007/978-3-030-93148-3_4

imperfect ways. Olsen (2012) explains the misrepresentation possibility in Part 7 (pages 203–217), and a longer summary is in Archer, et al. (1998). Since observation is not the same as keeping records, realists distinguish the errors that can arise at the record-keeping stage from perception and observation itself. If mis-representations are possible in data, then claims about the real world, arising from looking at data, might also have misstatements. Thus realists often argue for returning to gather more evidence about the real world and trying to re-examine what actually happened, in order to get more scientific and more refined findings in a fresh research study.

Realism has been used in social science as an alternative to a constructivist ontology. Porter, while defining and explaining both realism and critical realism, says 'realism can be contrasted with constructionism, which holds that society is constructed by individual subjectivities' (2011: 256). Critical realism, he says, is a more narrow form of realism which specifically argues that structural features arise out of human social development combined with nature, and that change over time is influenced by emergent features—which are often unintended features—of the open systems that interact with each other. When we think of discourse, language habits, and class structures we can see that realism is not just an assertion that 'social class exists', but also the assertion that agents influence class and class influences agency. Realism is also different from phenomenology, if the latter is taken to refer to a metaphysical stance that phenomena are always studied from within and therefore do not exist in a pre-fashioned way prior to the researcher's investigation.

▶   Defining Retroduction: Open and Closed Retroduction was introduced by realists mainly as an alternative to induction (Bhaskar 1997). It was also meant as an antidote to excessive use of deductive thinking in social, transformative forms of research. The view of Bhaskar was that research is inherently transformative, whether or not the agents doing it are recognising that. Reflexivity and retroduction thus arose as a way for self-aware agents to hold dialogues and carry out research, without relying upon either simple induction or simple deduction for the validity of their findings. Here is a short definition of retroduction from Olsen (2012: 215–216). There are three forms retroduction takes.

First, we can ask, what has caused the original data to show the patterns it showed? Retroduction would involve inference to the best answer to this question. It could involve reviewing why did someone start off with the topic that they had set up in the beginning? Were there theoretical reasons? Or was it based on principled ethical reasons or material interests? Was the promoter or funder an interest group that is powerful or marginalised? Why did respondents answer questions in the way they did?

Second, retroduction would ask what causes unique, unusual, deviant, or inexplicable cases to turn out as they did. New theories or a wider, more encompassing theory might help to explain a wider range of cases. Another aspect of this second prong is making a fresh typology of cases. Third, we can ask, what are the complex standpoints and viewpoints that are behind the contradictory results in this research? There is danger of a project widening out too much, so narrower aspects of retroduction may be needed for one specific project. Doing retroduction offers opportunities for creative problem-solving, dialogues among opposed stakeholders, and new findings. In Chap. 8 I have provided examples of empirical applications of the retroductive logic.

In this book I explain that 'closed retroduction' can refer to using additional variables in a regression model, from amongst existing data. This form of regression can ask or answer new questions and not just test the pre-set hypotheses. With qualitative data, closed retroduction may involve returning to the interview texts or videos to address questions repeatedly, seeing the range of responses. On the other hand, 'open retroduction' would refer to a more open enquiry in which we go outside the box of given evidence, moving to new rounds of expert interviews, observation, additional information, or other explorations that help to answer the retroductive questions.

4.1   Multiple Logics in Statistical Research: Some Exemplars

▶   **Tip** Many people believe that deductive logic and statistical regression models fit together with hypothesis testing. However, reviewing several statistical research projects suggests that a wider range of logics are being used for analytical purposes on a regular basis.

I will begin by giving examples of the use of the induction, ve deductive, and retroductive logics in adding to an existing argument. Usually in statistics the researcher thinks they are falsifying hypotheses.[1] This is the tradition that fits elements of

---

[1] Readers may realise that it is hard to specify what counts as 'statistics' or 'social statistics' because the nature of these subjects keeps changing. There is also much diversity, so these are schools of thought and not groups of people. I will count as statistics both using survey data and combining algebraic models with probability reasoning. The phrase 'social statistics' refers to the application of statistics in contexts that include social phenomena.

Popper's critical rationalism alongside elements of empirical (naïve) realism. (There are many other approaches in statistical interpretation, but I focus upon this one as it is dominant in textbooks.) See also Bhaskar (1979) who discusses types of naive realism.

Review the literature
Build hypotheses
Check existing data, enumerate cases, and gather variables
Test hypotheses one by one, or all together and then separately
Reject those hypotheses not supported by these data
Show the findings to others.

Implicitly this research design gives a tacit approval to the validity of the non-rejected hypotheses. Usually the null hypotheses get rejected, while the 'alternative hypotheses' are the interesting claims. The researchers use the following logics in developing their tests:

(a) They use deductive reasoning to set up the hypotheses. Deduction involves general principles based on existing literature, then making claims derived therefrom about individual units or particulars. The phrase 'deductive' is used to indicate that the conclusions do not rest on personal biases, suppositions, speculation, or emotive insertions, but merely upon the existing knowledge.

(b) Then they use induction to move from the findings to the conclusions. These studies use empirical data so they have a clear scope and boundaries in space and time. Within that, the authors generalise to the population from which the sample was taken. This generalisation is inductively valid, based on the evidence and the underlying reality.

(c) They use synthesis to present all this in a logical argument. Their synthesis often presents the review of literature and hypotheses as value-neutral elements. Any puzzles which arise from the incompleteness of the review, or any contradictions that arise, are presented as requiring new hypotheses. Once the initial tests have been run, the conclusions are drawn up and an agenda for future research is set out.

(d) Many people say the scientific method of hypothesis testing is deductive but that is an inadequate description of it. Retroduction is used in moving from a simple model to a more complex model with more parameters. Closed retroduction occurs when they choose new variables for the X side. Some studies move to 'multi-level modelling' to allow multiple levels of causal factor, for example regions, schools, and classrooms.

(e) Open retroduction may also take place. They can refer to experts, visit a fieldwork site, or consult newspapers and qualitative secondary data. If they use 2–3 data sets, then they must retroduce to move towards overall conclusions. 'What do I need from Data Set 2 to complement the gap in Data Set 1?' is a retroductive question.

One example where regression was used in a selectivity-adjusted Heckman formulation focuses on the odds of Y happening (a person working for pay), represented as a non-zero variable. These odds are measured with a logistic model called λ_i for each person i. The level of wages of the person is estimated to be higher if Lambda reflects a high probability of working. This typical method is called an 'econometric' method, since Lambda and Y are economic measures. Examples that use such methods include Kanjilal-Bhaduri and Pastore (2018); Olsen and Mehta (2006a).

Below, I offer a short overview of Kanjilal-Bhaduri and Pastore (2018) and three other exemplars that use statistical methods. They varied considerably in their use of mixed methods, ranging from survey research to action research.

Appendix Table 4.3 shows the details of variables, outcomes, and causal pathways for each.

▶   **Statistical Exemplars Involving Pluralist Theorising**

Kanjilal-Bhaduri and Pastore (2018); Olsen and Mehta (2006b)

- Topic: Minority Ethnic Discrimination in Wage-Setting
- Place: India
- Scope: Thousands of randomly chosen cases
- Methods: Regression with Two Dependent Variables, Structural Equation, Heckman

Kelly et al. (2011)

- Topic: Testing the Effects of an Autonomous-Working Treatment in Workplaces
- Place: USA
- Scope: Hundreds of workers in control and treated groups
- Methods: a Quasi-Experimental Controlled Trial

Abou-Ali, El-Azony, El-Laithy, Haughton and Khandker (2010)

- Topic: Factors Affecting Poverty Via Sanitation
- Place: Egypt
- Scope: Thousands of randomly chosen cases
- Methods: Randomised Controlled Trial of a Social Treatment

Mukherjee, Dipa (2012)

- Topic: Harmful Child Labour
- Place: India
- Scope: Thousands of children
- Methods: Secondary Statistical Analysis with Closed Retroduction

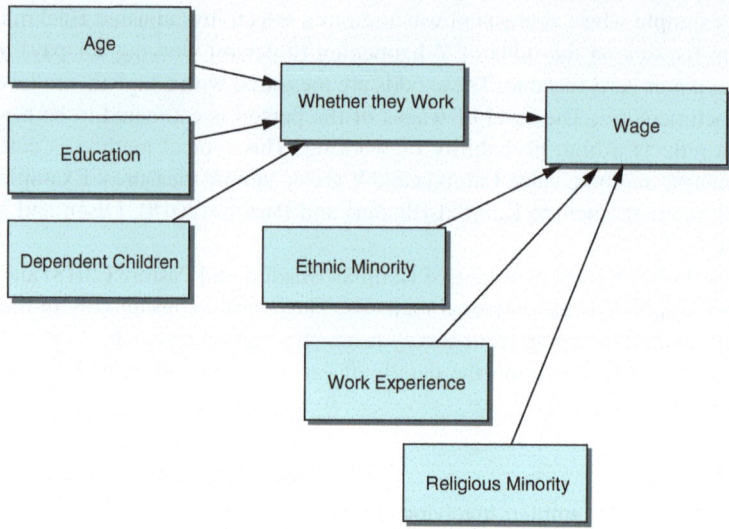

Fig. 4.1 Typical analysis of the labour supply decision and wage outcomes. (Sources: Kanjilal-Bhaduri and Pastore (2018); Olsen and Mehta (2006a, b))

Figure 4.1 sets out the two-step causality involved in the first exemplar.

Equations 4.1 and 4.2 denote the kind of selectivity-adjusted (named after Heckman) model used here. This classic model in western labour market analysis has been estimated hundreds of times in the published literature. In the equations W refers to working for pay (yes or no) and Y for wages (amount per day or per hour).

$$\text{Logit}\big(\text{odds}(W)\big) = f\big(\text{child, region, single}\big) \tag{4.1}$$

$$Y = g\big(\lambda, \text{educ, region}\big) \tag{4.2}$$

So far, λ represents the odds of not working; W is the variable 'they worked for pay', measured 0 for no and 1 for yes; 'child' indicates whether the respondent has dependent children; 'region' indicates which region they live in; 'single' indicates marital status; and 'logit' is a function that transforms the 0/1 outcome (works/doesn't) into a workable linear regression equation by taking the odds of the W variable (Rabe-Hesketh and Everitt 2009).

Next, Eq. 4.2 outcome, Y, reflects the wages of the workers. Those who worked have data for Y and those without paid work have an estimated (predicted) wage rate, but no data. Instead, however, they do get assigned a 'lambda' reflecting the probability or odds of them having/not-having any paid work. This lambda enters into the second equation. For women it is usually a negative element in countries like the UK, but in India the results are different. The two regions have very different market structures as well as cultures of gender (Table 4.1).

Table 4.1 Labour supply affects wage estimates in India (structural equation model)

Independent variables for wage equation	Male		Female	
	Urban	Rural	Urban	Rural
Intercept	6.979[a]	6.165[a]	6.794[a]	5.738[a]
Potential work experience	−0.023[a]	−0.006[a]	−0.012[a]	−0.002
Square of potential work experience	0.000[a]	0.000[a]	0.000[a]	0.000[a]
Education	**0.170**[a]	**0.128**[a]	**0.169**[a]	**0.129**[a]
Technical education	*0.022*[a]	*0.055*[a]	*0.007*[a]	*0.038*[a]
Lambda	***−0.0230***[a]	***0.171***[a]	***−0.095***[a]	***0.449***[a]

[a]Statistically significant at the 1% level. Potential work experience is an adapted age-related measure. Lambda arising from the labour supply equation measures the risk of **not** participating in the labour market. The technical term for Lambda is the inverse Mills ratio
Source: Kanjilal-Bhaduri and Pastore (2018: 527). **Bold = a factor deserving stress**. *Italic = a gender-differentiated result*

The letters f and g represent a simplification of the functions. In Eq. 4.2 for example we can have a sum of three terms and thus three coefficients: $\beta1$, $\beta2$, and $\beta3$ measure the linear model of the wage-rate predictions, based on three variables: lambda, education, and region; and we can add a discrimination estimator using 'minority status' (0/1) as a 4th variable in the g function, not present in the first stage f function.

There are some key differences between statistical equations like this and the fuzzy set approach I displayed earlier in Chap. 2. I will explain these differences in Chap. 6, because confusion can arise from having too superficial an approach to either of these methods. Leave this aside for the moment. I could add an original third equation if I were preparing to do new, path-breaking research. Equation 4.3 reflects factors that contribute to wellbeing, on a personal basis. Equation 4.3 offers another form of open retroduction: opening up a new area of related research.

$$WB = h\left(Y, \text{child}, \text{educ}\right) \tag{4.3}$$

In this third equation, I suggest that working hours Y affects wellbeing 'WB'. I used the letter h to indicate an equation with some function form, again perhaps linear sum of terms. 'Child' and 'educ' are the same variables as before.

The three equations create a structural equation model. Working for pay increases the wellbeing of the respondent and Eq. 4.3 makes this explicit, thus enabling it to be a tested proposition. Both the Popperian hypothesis-testing approach and SMMR tend to promote discovering new testable propositions (Downward and Mearman 2007).

One example of the initial model is in Kanjilal-Bhaduri and Pastore (2018) who looked at earnings in India, showing two gender-related findings: firstly, the returns to formal education are similar for men and women, yet technical education shows gains for men exceeding those of women. Secondly, the model shows rising wages over the age of worker, all else being held equal.

Remember that here 'lambda' represents wages that would have been obtained among both those who were and those who were not in the paid labour market. Lambda reflects the risk of being selected out of the labour market, that is jobless.

Lambda itself is non-negative, since it is a probability. In western formal labour markets, this normally has a negative coefficient in the wage equation. Thus, those who are less likely to work for pay have lower wage estimates and less of lost earnings, all else being held equal. However, in India this pattern is only observed for urban residents (Columns 1 and 3). For rural men and women, the element is positive. This means those excluded from the labour market would have earned more if they had been able to get paid work. The author hardly even discussed this intriguing result. If we stick to deductive arguments and avoid deviating from the original hypotheses, we may miss an important finding.

Such findings can form the basis for interesting further enquiries. My own current research has moved towards making an explicit wellbeing estimate after the wage estimates. Others might explore wellbeing qualitatively. Many other possibilities exist for combining the regressions of Eqs. 4.1 and 4.2 with qualitative data.

Kanjilal-Bhaduri and Pastore (2018) did not engage in a mixed-methods study nor did they use triangulation in this article. Their logic was

Review the literature—apply it with India data—fit the model—test hypotheses—conclude.

It was fundamentally a deductivist article. Notice (Table 4.1) that education has a positive association with higher wages inside all four groups. Technical education caused a much higher rise in men's than in women's earnings. Again, these firm findings received little discussion and are worth reflecting on in a broader context of gender and pay-gap debates.

The paper illustrates how the standard 'regression-methods based paper' is kept tightly focused on its initial research question. Using systematic mixed-methods research it could range more widely. It may not stay focused on a single RQ but instead broach new questions as part of its findings. Thus, quantitative methods may be cautious and avoid risks, while SMMR may take risks and be more innovative. It is still possible to be rigorous, and transparent and sophisticated, either way.

Suppose we want to extend this wage study into a broader analysis of how wages and work affect, or are associated with, human wellbeing. A lot of further work has to be done. Some of it can be based partly on survey research. For example we often measure wellbeing using a classic indicator of how satisfied a person is—their answer to this question: 'How satisfied are you with your life this month?—Very satisfied, satisfied, neutral, dissatisfied, or very dissatisfied'. Many variants on this simple measure have been proposed. This form of measure, known as the life satisfaction approach to happiness, or to wellbeing, is widely used. It has a Likert scale as the answer options. It could have a longer scale, 0–10 for example.

To combine mixed methods with this model is very easy, but very costly. First, we have to examine the scope of the research question and what new, original question we want to answer. Depending on that assessment, we might do three focus groups in one locality, or ten survey questionnaires in each of the ten localities, to supplement some survey data that we may have on a wider basis.

Eq. No.: Dependent Variable	Independent Variables
1: Has a Paid Job	Number of Dependent Children, Region, Age & Age^2
2: Wage Rate (£/Hour)	**Has a Paid Job**, Marital status, Education, Work Experience (Years)
3: Well-being (Scale)	Hours worked, **Wage-rate**, Education, Region

Fig. 4.2 Typical variables in a multiple equation model

Alternatively, we could use triangulation to add this data set to another, larger data set on wages. Perhaps we have access to free, secondary data from the Labour Force Survey (e.g. in India, the Employment and Unemployment Survey held by the National Sample Survey Organisation). We can be ambitious in combining two data sets, as long as they are both random sample surveys.

To use secondary survey data you would need all these variables in a single data set (Fig. 4.2).

If we include men, then we need a gender variable. We can then interact gender with other variables to get the sex difference in causality to stand out. For wellbeing, too, gender differences are considerable. As a practice run, one could draw a path diagram showing the Heckman two-step wage model plus the wellbeing outcome. This task involves labelling the wage as Y_2 and the wellbeing index as Y_3. It has many paths, involving both moderation (which means interactions of effects) and mediation (which means indirect and direct pathways).

This study can easily go forward but what elements of logic will you be using? Again it depends on the research question. Three illustrations are offered here.

(a) Do some hypothesis testing (induction), then go out to the field, gather semi-structured interviews (open retroduction), conclude about that locality (induction; some call this analytical generalisation), bring the results together (synthesis), draw broader conclusions (synthesis), ask whether the analysis has a good fit and if not, what needs to be added (retroduction), and add these and re-run everything (backwards variable removal with inductive hypothesis testing). Or:

(b) Start with three focus groups in one area (open, semi-structured research methods leading towards induction), hold these data pending triangulation, then look at the secondary data for a region that includes this area, develop an explanatory model from the data (retroduction and deduction are used together), retroduce about what outcomes really matter to people in their wellbeing (retroduction and then synthesis), reassess the relevance of both sets of data—qualitative and quantitative—to this research question as re-phrased, and finally start some hypothesis tests with different geographic scopes (deductive), and conclude about your local area (synthesising). Or:

(c) Start with the problem to be addressed, gather up to ten stakeholders, develop an action plan (action research, see McNiff and Whitehead 2011, for step-by-step advice), try out the actions, start making data records to do an ongoing evaluation of your joint experiences (prepare for induction), gather up the secondary survey data, build the model (closed retroduction), and test hypotheses on requests from the stakeholders (deductive), do a survey of the stakeholders and some customers or users, draw up those findings (inductive), bring the two together (Triangulation), hold a discussion meeting with stakeholders (which may involve open retroduction, but also may involve presenting earlier findings), conclude the evaluation work (a synthesising moment), and bring the whole thing to a close (a synthesising moment).

Other options may occur to you. Deduction plays a significant but relatively small role. Thus, the use of quantitative data has been established here as a supplement to the whole research enterprise and not a dominant activity. Deduction is not a paradigm.

Some textbooks offer a rather different view. In trying to be helpful, they recommend a clear, tight, narrow research question. Perhaps to keep projects within budget, they omit all the options for exciting research which get you out of your desk, moving in social environments observing, recording, and setting up learning situations. This is not the only way to do science. I urge that we consider doing things in a more complex way to encourage career learning.

Flyvbjerg (2011) showed that the learning environment is a key step in facilitating social learning. The learning environment is set up within a research project. This situation needs good communication and multiple voices, and it is not just deductive. There will be, within that environment, many logics that we use to move from point to point in our research design and in developing the findings.

The most important gap in what I have said so far about the wage-rate regression studies is a risk of lack of data. If you want to study wellbeing, it may not be covered in one of the data sets. Indeed the measurement of wellbeing is itself fraught with debates. One measure uses 12 Likert scale questions. It is known as the General Health Questionnaire (GHQ) and we found that it was more successful to break this up into two parts and analyse them separately, rather than analyse them all within one index (Bayliss et al. 2015). Other scholars too have debated this question. In India the labour force datasets have no measure of wellbeing. Researchers are actively looking at how to combine data sets on wellbeing outcomes with those from other areas. A mixed-methods project could offer a synthesising solution.

Using mixed methods we can suggest institutional ethnography for a study of wellbeing of workers. The ethnographer visits a workplace with full permission and in full view of employees, with informed consent and usually posting up notices to indicate the topic of research. Excellent institutional ethnography will not be on a large scale, but will be very in-depth. There is a tradeoff: if you do the ethnography only in one locality, you cannot generalise nationally, but if you do it nationally you won't have much time for each ethnography. It is wise to be very honest and open about the breadth versus depth of the ethnography. In writing up findings, we must state precisely how much time was spent by which researcher in which place.

Institutional ethnography may enrich the study very well. Using methods of observation, unstructured interviewing, reviewing documents, and then discourse analysis, the research team will bring together synthesised findings from the ethnography and discuss these in detail. A computer database will hold most of the recorded findings. There will be sophistication and transparency in how evidence contributes to the main conclusions.

On the other hand, the use of institutional ethnography presupposes informed consent and indeed the consent of employers so that researcher could go on in a work site. This might not be possible in a sector that has precarious workers, or high informality, or in Indian settings, so there might be a need for other methods such as using semi-structured interviews at the workers' homes, or oral history or other techniques.

After carrying out institutional ethnography (or other method), you create bridges from the statistical study to the ethnography and back. The timing is a decision you make as you go along. Creswell and Plano Clark say more about the sequencing of the study elements. The study by Kelly et al. (2011) illustrates institutional ethnography combined with a serious survey including randomised control groups. However, as the survey did not have 100% reply rates, the ethnography helped a lot in ensuring reasonable conclusions from the targeted control and treated departments.

4.2   An Exemplar Using Participatory Research with Panel Data

▶   Tip If a research project has 'action research' in its design then it will have participatory activities. As a result the study is not 'naturalistic', that is the agents in the scene are not left alone. It can still lead to a wide range of learning outcomes. This learning happens during the project, and not just after the project is done. Another reason for multiple learning innovations is that several types of stakeholder all stand to gain from the project over time.

Kelly et al. (2011) developed a complex regression equation based on extensive research in one firm. Results-Only Workplace Environments (ROWE) were tried out in a variety of departments in that firm during this mixed-methods action-research project. The panel data for statistical analysis brought pay data from inside the firm together with sample surveys applied as a pre- and post-test among those who were trying out the ROWE system. This ROWE system basically promotes workplace autonomy. Besides the departments with Results-Only Workplace Environments, other departments acted as control groups. Discussions could take place among all the various people at any time.

Two hypotheses were put forward. A rather obvious one was that the workers with ROWE would have more control over their schedules; it was a form of flexible working and this was supported by the findings. The second hypothesis was about what results schedule control would cause. The authors explored how work-family conflict of various kinds rose or fell between pre- and post-test among the workers affected. The quasi-experimental design is helpful in allowing the analyst to compare like with like. Some families might have lower work-family conflict for the direct reason that dependent children become older and more independent, for example (these direct lines of causality are not shown in Fig. 4.3). The test of

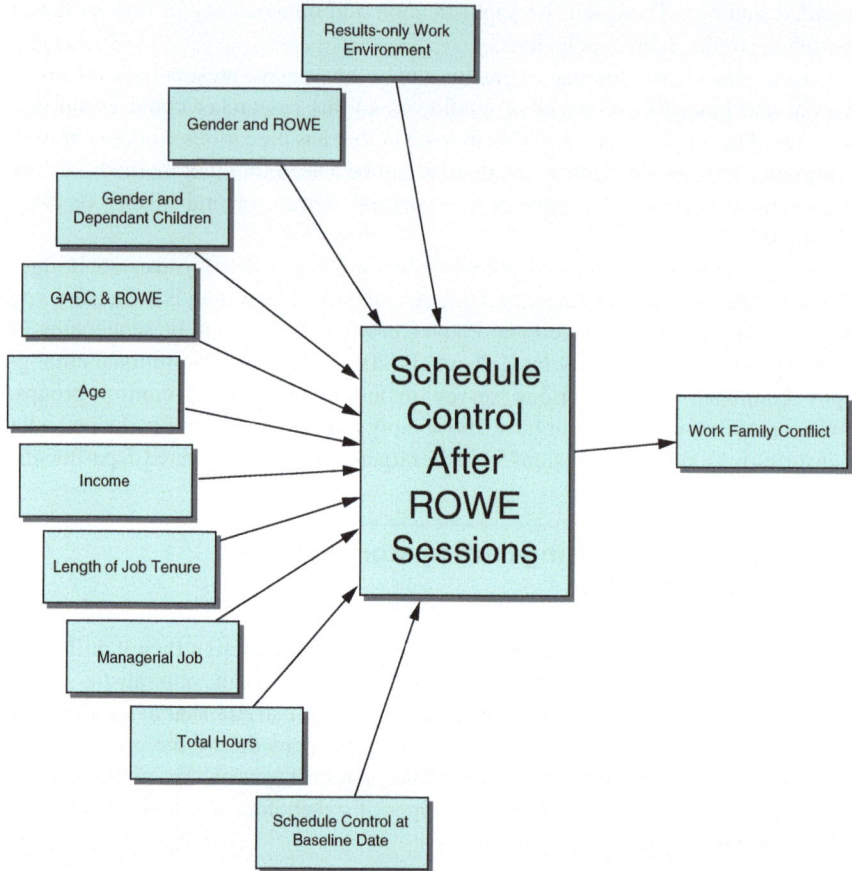

Fig. 4.3 Schedule control, a mixed-methods study with quasi-experimental statistical panel data analysis (Source: Kelly et al. 2011)

whether schedule control improved the work-family balance, reducing conflict, was a key outcome of the study.

This study shows that it is possible to set up a mixed-methods statistical study. The paper had three authors, whilst many other people were involved in the workplace project itself. The action-research elements would have had effects during the study. They did not wait for hypothesis testing before moving into discussions of what might work better, either at work or at home. Learning was assumed to be going on. The logic of this study is complex, with key aspects as shown below.

Review of Literature—Deliberation About Intervention—Plan the Intervention
Plan the Pre-test and Retest Survey—Interviews—Observations
Divide the Workplace into Contrasting Units, Reach Agreement
Conduct the Survey Before and After ROWE Intervention
Carry out Post-Intervention Contrastive Explanatory Modelling

Closed Retroduction by Adding Variables to the Model
Open Retroduction in Workplace and in Homes

The study had one key difference with the standard 'action research' model. In 'Action Research', the intervention itself would be changed over time and thus is seen as non-rigid (McNiff and Whitehead 2011). Stakeholders would have influenced it. However, in the Kelly et al. study, the participatory discussions seemed to not affect the basic ROWE situation in the treated units. These departments carried on working as Results-Only Workplace Environments. It is likely that if the ROWE were less popular or visibly harmful, they would have stopped this trial. Thus, the project has some elements in common with randomised control trials (RCTs), too.

This research design is also strongly similar to a Qualitative Comparative Analysis (QCA) study. I will draw out similarities in Chap. 6.

In the Appendix I have listed the variables used here. The ROWE itself was an event, involving a change of structures and institutions at the workplace. Instead of working set hours each week, the ROWE worker can decide their own hours and weekly pattern, being judged only on results and deliverables. This study illustrates the way a structural feature of a social institution can be posited as a 'variable', with some success. For sociologists this paper offers a way to operationalise social institutions very carefully and trace change over time. However, the paper worked well, because of the discretisation of institutional change into a simple contrast: pre-ROWE and ROWE situations. This obviously is a great simplification and may have proved rather challenging in practice. If the ROWE system had been self-evidently a boon, it might have been hard to keep the non-ROWE units away from this new set of norms.

4.3   A Statistical Exemplar with a Randomised Control Trial for a Social Intervention

As a third illustration of regression, I now turn to a paper by Abou-Ali, El-Azony, El-Laithy, Haughton and Khandker (2010; see Fig. 4.4). They examine whether a national social fund scheme had a substantial impact by reducing poverty in Egypt. The treatment is designated as the social fund for development, also denoted as Mixed SFD Treatment in Fig. 4.4.

The social fund for development (SFD) is a long-term development scheme which provides funding for initiatives to help in various areas of economic and social development. It involves participatory management, enterprise, or grassroots initiation of activities such as new forms of sanitation. Social funds were applied to support the development of small enterprises, microcredit, and community infrastructure (Abou-Ali et al. 2010: 522). The SFD funding was distal, with each project being implemented at the proximate end by other agencies. Each implementation involved either an investment or the delivery of a service. I have used the term 'treatment' in Fig. 4.4 because the randomised control trial (RCT) method used here is often seen as analogous to evidence-based RCTs in medicine. A common question

Fig. 4.4 Egyptian treatment of localities using a social fund for development (SFD) (Source: Abou-Ali, El-Azony, El-Laithy, Haughton and Khandker (2010: 531, focused on potable water projects))

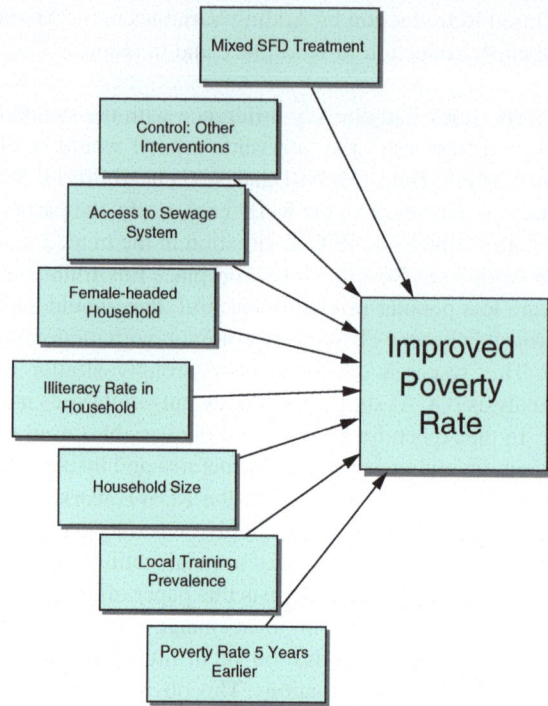

about treatments is how much of the treatment to administer. Quantification is seen as very important in such situations. Here the treatment is measured in units of money.

The randomised control trial in the paper had one treated group and three control groups. The regression takes the form of a treatment model without interaction effects. The unit of analysis was not individual households, but rather data arising from villages and wards (N = 1290). The control groups vary in the amount of other (competing) interventions received in the chosen areas. Within Egypt, some areas had non-SFD funding interventions. Having noted these, it was possible to compare SFD, non-SFD, SFD with non-SFD, and no-intervention areas. A small geographic unit was used. This enabled the team to consider each treated group together, even though each consisted of units which at the spatial level were not always geographically contiguous.

The success of the schemes was measured on various dimensions, including sanitation, poverty reduction overall, and income generation. The means of improving people's lives included a variety of initiatives in education, health, potable water, as well as economic support for enterprise development. I will focus on the improved poverty rate as a key outcome. A poverty headcount is a count of the number of people living below the agreed poverty line. This headcount rate was positively affected (i.e. lower) in areas with the interventions. The study covered the 1991–2007

Table 4.2 Success of scheme for poverty reduction (single regression) and then, below it, we get a note, Abou-Ali, et al. 2010: 522

Dependent variable: headcount poverty rate (%) Unit of analysis: 1290 wards in villages, aggregate data	Coefficient	*p*-value
SFD piped water project in place (1 = yes, 0 = no)	*−0.81*	*0.00*
Households with access to a sewage system in 2004/2005 (%)	−1.53	0.00
Average household size in 2004/2005	0.26	0.07
Female-headed households in 2004/2005 (% of all households)	1.65	0.08
PSU had one or more training programmes (yes = 1) in 2004/2005	−0.16	0.64
Illiteracy rate in 2004/2005 (% illiterate among those older than 10)	−0.45	0.01
Headcount poverty rate in 2000 (%)	0.07	0.00
Female-headed households in 2000 (% of all households)	0.06	0.02
Intercept	1.08	0.17

Note: Observations are weighted; propensity score weights are used; the adjusted R^2 is 0.168, meaning 17% of the variance in poverty rates was explained. The data come from 1290 primary sampling units (PSUs), each being a ward in a village
Key: The table contains slope estimates from regression showing the impact of SFD potable water projects on the headcount poverty rate, comparing PSUs having 'mixed' SFD and other interventions (C) with PSUs having only non-SFD interventions (E). The p-value is the probability of being wrong in rejecting the null hypothesis of no effect for a given explanatory factor
Source: Abou-Ali et al. 2010: 522. The original source is those authors' calculations based on HIECS and community survey 2004/2005 and 2000 census

period of intervention. I will focus on SFD piped water projects (the paper goes on to look at each area separately: sanitation, health, education, etc.; Abou-Ali et al. 2010: 532). Overall the statistical model is a cross-sectional regression using data from both 2000 and 2004/2005. The later year is used for outcomes and both years are used for some independent variables.

The factors considered to influence the success of the piped water intervention are shown in Table 4.2, including whether they had a sewage system in place in 2004/2005. Those with a sewage system had less poverty overall and this acts as a control, removing this 'effect' from the estimated impact of the piped water intervention, which was associated with a further reduction in the poverty headcount outcome. Other factors included household size and being female-headed, the availability of local training programmes, illiteracy, and being female-headed five years before.

Time was treated carefully to allow the short-term (five-year) impact to stand out. In this regression, the level of poverty in the initial year was adjusted for by using the 2000 poverty headcount figure as an independent variable (see Fig. 4.1 and row 1 of Table 4.2). Effectively this creates a structural equation model of change over time in poverty. It gives 'results' cross-sectionally for the 'impact' of the interventions, after allowing for the whole historical burden of other past interventions and circumstances by putting 'poverty headcount 2000' in the regression as an explanatory variable for 'poverty headcount 2004/2005'.

This ambitious article thus paves the way for further analysis of individuals and households who might be affected differently from the aggregate poverty impact in their location.

Queries have been raised about the randomised control trial method in social settings. Mixed-methods approaches are incompatible with RCT Impact Assessment. If need be, in social contexts, I would tend to sacrifice the RCT component, not the use of mixed methods (Olsen 2019, *EJDR*). In Abou-Ali's case (2011), the authors did not claim to have done any research of a qualitative kind nor did they engage in action research. The issues I would raise are the ethical matter of waiting for the RCT trial to be complete in its post-test stage, the sample size requirements, and the deductivism of RCT logic.

It would be unethical if the RCT is set up a priori in 2000 or even in 1991 and we were made to wait till the publication date 2010 to find out where best to invest money (social fund or not SF; and within the SF, which areas are most efficacious). If we interpret the RCT as a deductivist hypothesis-testing exercise we may find there is too long a delay between results and feedback. There is often a staggering of the rollout of the intervention, and so for a successful treatment rumours may emerge leading to an unethical or biased choice of the second-stage treated groups. There could even be bribery to entice the treatment into one social area. The sample size required for statistical power would be higher and higher as we think of smaller and smaller units for a decision; yet if the first test on 30 units shows a poor result, then we need to think as follows: is the failure for deep reasons that involve widespread social or physical conditions, or structural features common to a wide area? Or is the failure due to some narrower factor? Only in the latter case would the rollout to another 30, 60, or 90 more cases be justifiable. This is a question in logic and it requires evidence and discussion, not just more cases. Finally, I had doubts about the pure deductivism of the use of survey data alone. Instead it can positively be a benefit to add interviews and observation to the research design.

Thus, ethical issues surround what happens during the long period of time from research design to research publications.

An alternative way to see the situation is teamwork rooted in the actors and agents of the policy and practitioner world. In such a teamwork project, aims and objectives are under discussion over time, sampling might be altered midway through, and preliminary findings would be taken seriously as the basis for retroduction. My model of how this research would be improved with mixed methods might suggest these stages:

Review of literature—data collection begins—induction—
Interviewing—voices in meetings—open retroduction—
Revise the intervention, and the survey data content—
Structural model of final outcome levels—closed retroduction.
Time →

The differences would be vast; the statistical model would no longer take the same mathematical format of assuming unchanged baseline levels nor focus on change in a single indicator of success. Instead we may introduce changes in the baseline levels and an alteration in the indicators of success. Empowerment itself might be one of the latent factors whose improvement could be measured. Both qualitative evidence and survey evidence could be combined. Giving power and voice to the local research team, to the local policy practitioners, or to political agents would be a divergence from the RCT norms. It is possible to argue that the RCT tradition involves an assumption of a fairly conservative, slow-changing society with fixed agendas for change, allowing slow measurement over long period of time in harmonised surveys. The RCT tradition might also be considered deductivist, although there is no necessary reason why that must be so. Perhaps the training of the mathematical experts who set up the RCT regressions tends to encourage deductive skills and discourage inductive and retroductive skills. Nevertheless, there is overlap of the mixed-methods approach and what actually is reported in this article. Abou-Ali et al. openly mention aspects of the society that helped sanitation to perform in one way but education in other ways, not tightly tied together. The article also shows a huge baseline knowledge of what development has tried to achieve in Egypt. However, as presently written the article is entirely based on quantitative data and has only very poorly theorised the development process itself. It uses no qualitative data.

My last regression exemplar studies children working in India. Mukherjee (2012) analysed child labour as a harmful outcome of family joint decision-making. Here, children working for remuneration find this work competing with or even obstructing the child going to school. The decision to go to school could co-exist with the child doing some paid work and this would be considered a harmless combination. In this literature, the specific phrase 'child labour' is used to refer to harmful forms or amounts of child work. A clear set of definitions is offered to set the stage for a measurement process, using survey data, and then the testing of regression-based hypotheses. The similarities with the RCT study are clear at the level of logic:

Conceptual development—data collection from a secondary source—
    Closed retroduction to see if explanatory power is gained from using new variables—
        Structural model of harmful child labour levels—closed retroduction
    Time →

The Mukherjee study illustrates how a study using regression and hypothesis testing need not be purely deductive in its logic.

Mukherjee estimates the factors affecting the risk of harmful forms of child labour in India. She uses a multinomial model where going to school, working, or not having **either** a job or schooling are options. The multinomial model uses probability modelling to allow each outcome—among these three mutually exclusive options—to have not only a 'probability' assigned to it, but an explanatory model with all the independent variables. These explanatory models are estimated simultaneously, giving 3 k coefficients for three work options. In Fig. 4.3, I provide details of the child labour probability model, whilst the author also estimated the schooling outcome model and a third model for the likelihood of not being in employment, education, or training, known as 'NEET' status.

The addition of two extra outcome possibilities is a parallel logical structure to the use of four SMO outcomes by Cress and Snow (2000). We have here three parallel regressions, whilst Cress and Snow had four parallel QCA analyses (see also Chap. 6).

For an outcome whose probability we want to estimate, we can choose either a logistic, logit, or probit model. These are similar yet use different units for the measurement of probability latent variable outcome. The idea that the outcome is 'latent' is helpful, as it moves us further towards advanced forms of quantitative method. Mukherjee chose to use a logistic model. The outcome is measured on a logit scale using the logarithm of the odds of being in harmful child labour. The unit of analysis is the child.

The model has a yes/no binary for its dependent variable. The explanatory variables include being in an ethnic minority or religious minority because the different ethnic groups in India have varying experiences of both work and school. Here, the religious minority status is Muslim. To represent access to the means of production and hence farming, land owned is inserted as a variable. With over 70% of the population living in rural areas, this variable is a powerful predictor of children doing long work hours for pay or other remuneration. Some are paid 'in kind'. This refers to them receiving grain or cloth as payment for work, either daily or on a seasonal basis. Many children also work in their own families. This kind of farming work was allowed for by Mukherjee.

The main issue posed in Mukherjee's paper is whether the family decides to invest in education to maximise the child's later earning power and productivity, or whether they take children out of school to gain income in the very short term. By 'schooling' we may mean either attending school, or passing each grade level at school, for example passing the primary school final year or passing secondary school. Many young people attend but do not get the pass grade. This is meant to predict a lower probability of children working. If a child succeeds at school, then the investment in their education is seen as paying off. Ideally, we need separate measures of attendance and achievement.

Mukherjee (Fig. 4.5) inserted as further explanatory variables a proxy for wealth and a rural/urban indicator. Wealth of a household is proxied by the household's per capita spending per month. The greater the household's wealth (proxy), the less likely they would intend to have their child do harmful child labour. Having a child in school is seen, in this literature, as a 'luxury'. The rural/urban variable allows for these impacts of wealth and education to vary by location.

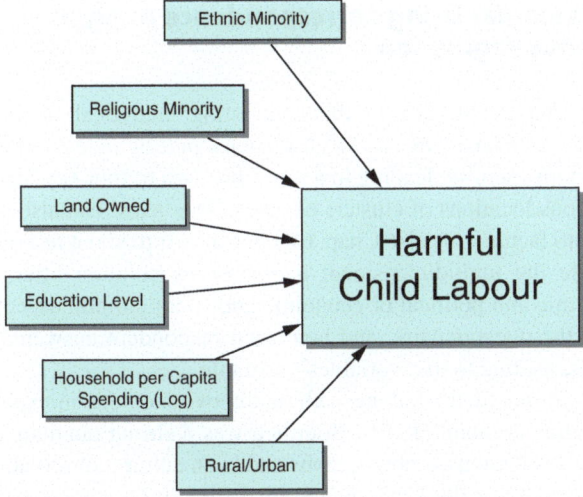

Fig. 4.5 Analysis of harmful child labour. (Source: Based on Mukherjee 2012)

4.4 Warranted Arguments and Two Caveats for Strategic Structuralism

So far, I showed that the refutation of hypotheses has a limited purchase, in social science, even though it has an important role to play. The hypothesis generation process is a social negotiation and is a key part of the research process itself. In using mixed methods we allow a variety of new theories and hypotheses to be assessed, before we start working on specific, well-defined, and well-operationalised hypothesis tests. Furthermore, after doing some of this kind of work (using the testing type of logic, often called deductive logic), we also want to carry out some retroduction and perhaps even some induction if there is time and if funds allow it. This iterative approach was advocated by Danermark et al. (2002), too.

A pluralist approach can represent two to three key arguments in a set of findings, so we do not put forward just one argument, but rather the pros and cons of a series of serious competing approaches.

▶ **Tip** If the research design includes gaining authentic 'voices' in the qualitative parts, whose is the voice when interpreting a regression model? The author's voice becomes one of the key voices. Therefore it is acceptable, within reason, to use the first person 'We' or 'I' in writing up the findings.

4.5   An Exemplar Using Correspondence Analysis Without Regression

It is a useful change to turn to the contrasting analytical method 'Multiple Correspondence Analysis' (MCA). MCA involves putting many variables together in a data-reduction exercise, leading to a set of just two to four key 'axes' that summarise the various locations of clusters of cases. The 'axes' can also act as indices and are similar to factors (see also Chap. 6). Each axis spread out respondents across its range, while the multidimensional nature of its solution offers distinctions between economic and political or economic and socio-cultural aspects of society. Depending on the questionnaire, and how each respondent answered each of the many questions, leading to the 'variables' available.

Bourdieu used this method along with qualitative methods in exploring French and other societies in detail (1979). Bourdieu was a strong qualitative researcher, who along with his team used photographs and keen cultural observation as well as the questionnaires. Over the years he and his teams did a whole series of mixed-methods research projects (e.g. Bourdieu 1995). First, I will explain MCA in brief and then discuss the logics and reasoning in Bourdieu's concluding arguments.

MCA was used by mixed-methods teams led by Bourdieu and others, to great effect (Bourdieu 1979; and see also Silva, Warde and Wright (2007) and Savage, Silva, Warde, Gayo-Cal and Wright (2009: Chapters 4–5)). In particular, as seen in the illustration in Fig. 4.6, the two axes most prominent in the reduced solution summarise all the variables and illustrate where cases are to be placed. It is crucial that the spread across the space is sufficient to separate groups or types of cases, and yet the similarities of the cases within these groups are sufficient to promote a concise summary. Some authors overprint a 'cloud' of individuals, just barely visible in Fig. 4.6, showing that the people themselves can be rated on each index. The 'cloud' concept is discussed more in Savage et al. (2009).

Figure 4.6 reflects 41 questionnaire questions, covering 7 fields of cultural activity (TV, films, reading, music, visual art, eating out, and sport). Thus, MCA is an efficient way to summarise a lot of information. MCA does not use causal or explanatory approaches to set up X–Y relations. It treats all the variables as equal players with equal weight, assigning value to each in relation to each 'index' or axis value; then it tests alternatives for the axis definitions until an optimal solution is found. Ultimately MCA is similar to a factor analysis with factor loadings and each case getting a 'factor score', but it uses all the information (as does principal components analysis); it avoids calculating any goodness of fit indices and it does not use the same formulas as factor analysis.

Instead of statistical fit indices, MCA is based on a set of functions dependent on the 'mass' or broadly the centrality of the data for each variable, and their correlations—hence the term correspondences. The 'mass' is not the Euclidean average. The method is suitable for all three types of quantification approach: binary, ordinal, or continuous measurement. A 'modality' or set of modal measures is set up for each measure, so that the information is all used: ordinal variables for example are usually broken up into a series of individual binaries. Thus, each can be mapped onto the diagram as seen in Fig. 4.6 *ex post*. In all, the figure uses 67 different

Axis 2

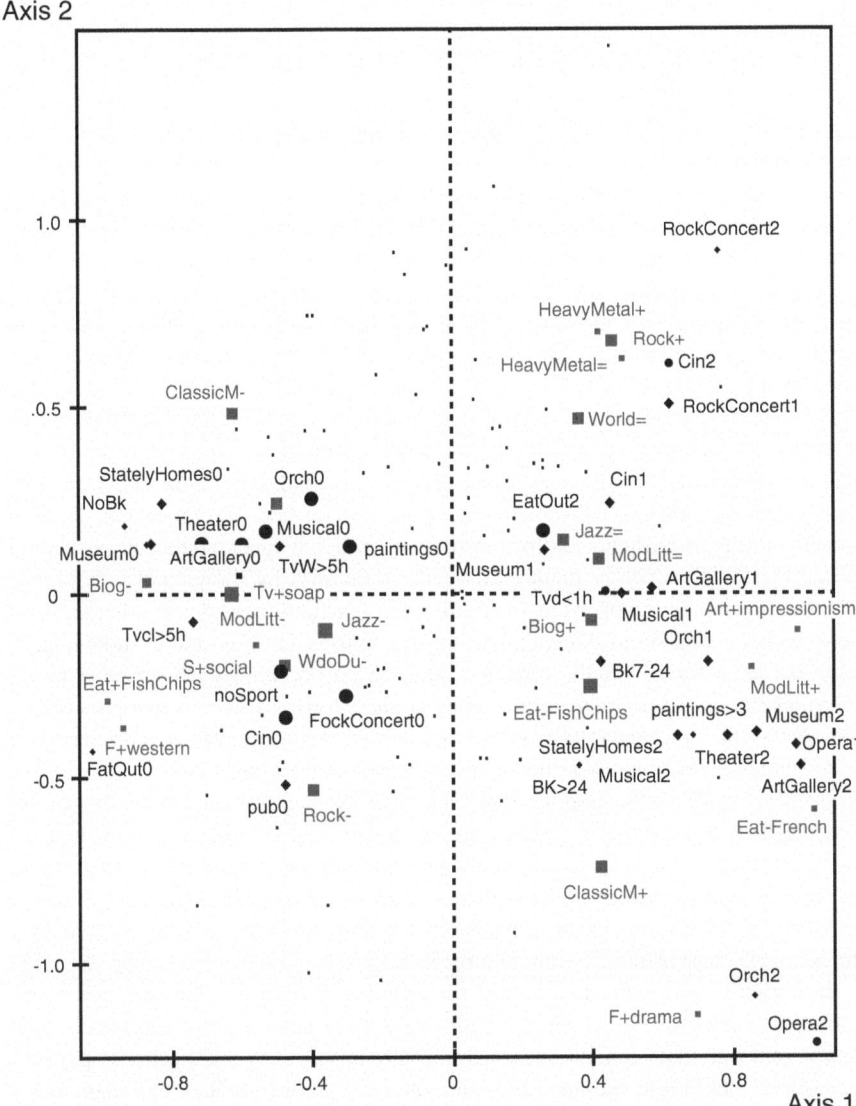

Fig. 4.6 Cultural capital (vertical) and a cultural omnivorousness index (Horizontal), 2003–2006, UK. (Source: Silva et al. 2007: 304. The label 'cultural omnivorousness' is provided in Silva's text at page 305, also called voraciousness or engagement in outside activities. The vertical axis is interpreted by me as involving so-called low culture (upper quadrants; called contemporary culture by Silva et al. 2007: 310) and high culture (low quadrants, called 'established culture' by Silva et al. 2007: 310))

modalities, showing that some of the 41 variables were ordinal and some were multinomial.

Using MCA, the teams studying the UK were able to reproduce similar results on cultural capital for the British society, based on theorising of culture and distinction similar to Bourdieu's (1979) approach. Both teams used semi-structured interviews intensively, too.

Bourdieu (1995) and Greenacre and Blasius (2006) also used multiple correspondence analysis (a quantitative method) with primary data of the survey kind. These works all use questionnaire surveys, but the sociologists combine these with interviews, observation, and other methods of research. In the Bourdieu (1979) case, the samples were non-random by social class, but random within class, while most other practitioners of MCA have used random representative samples (e.g. Silva et al. 2007).

The MCA method starts with cross-tabulations, which Bourdieu (1979) produces verbatim. He found very strong patterns when correlating engagement in cultural activities and types of eating-out alongside occupational social class. This was further patterned within broad social classes, such as farmers and managers. Thus as shown initially, Bourdieu's findings were structuralist: it appeared that class affected cultural taste. However, he nuanced the pattern findings by showing also that those who do not engage in upper-class activities still consider these types of activities to be of higher social status than popular, mass activities. He was able to show repeatedly that French individuals were in a bind, based on admiring some class-based activities but engaging only in those they could afford. To resolve this paradox he introduced the cultural capital concept as a competing concept to economic capital or resources. This theoretical turn led to a whole cultural turn in sociology (Silva et al. 2007) and was an important point of discovering the need for only 'weak' structuralism (my own term). Strong structuralism might expect deterministic causality, but weak structuralism would allow that many other factors, such as jealousy or being constrained from achieving what one wants, would lead to a lack of determination in the way activities or decisions are mapped onto (or arise from and on the basis of) class origin or economic resources.

Bourdieu's work was also famous for introducing into the debate a new term, habitus, referring to 'the structured structure' that structures and underpins human action. In rough terms, enculturated habit. Using this term he made sense of housing aspirations, eating-out jealousy, despising others' clothing choices, and many other social practices and other phenomena. His work had an enduring, world-wide impact.

The key move was to take up cultural patterns in a multidimensional space. He predicted that distinct structural and socially well-defined patterns do show up; then we can superimpose a separate structural or other variable as a 'supplementary' variable or as a post-hoc association. In Greenacre and Blasius (2006), the supplementary variables are not important in measurement, but are useful at the interpretation stage. MCA is a more expository, graphical, expressive method than regression, although they can overlap to some extent if regression predictions are graphed in smooth, helpful, and illustrative ways in 2D or 3D. We can easily show how patterns co-vary in the 2- or 3-dimensional space of the MCA (software SPAD built in French is useful for this; R and STATA also carry out MCA). A bivariate MCA has a two-dimensional space with factors 1 (horizontal axis) and 2 (vertical) being continuous estimated indices, based on underlying data. A multivariate MCA allows >2 dimensions to emerge and thus requires >1 2*2 diagrams to illustrate the XY, YZ, and XZ patterns.

In MCA, the underlying variables can be binaries, ordinal, or continuous. In SPSS, STATA, R, or other software we obtain the 'mass' or centre of gravity of each pair of variables and then the computer optimally sets up the indices, or factors or dimensions, so that each variable has a location in the space. This is not the usual 'mean' of the variable (which can be seen as a line at the mean, one for the X mean and one for the Y mean), but rather a mass index depending upon underlying weights assigned to a whole variety of observed variables.

To give a very simple example which was part of Bourdieu's original study, we could have a table of choices of preferred types of dining, by social class with occupational class in ordinal categories. The pattern of preference for each type of dining is a separate variable, so in total perhaps six variables. Then setting up MCA the first dimension comes out as closely related to economic strength for each class, and hence frequency of eating out. Meanwhile the second dimension that comes out will be a socio-cultural elite dimension. This is more likely to show which kind of eating-out they did: fish-and-chips among low-income groups and French restaurants among those who have more resources. Whilst they are correlated, the method of MCA is able to discern them, given good variables as inputs. Figure 4.6 was chosen to reflect the multi-disciplinarity of this non-causal, associative method.

It is impressive how strong a data-reduction the MCA can give. Whole sets of preferences about how we socialise, what art we like, what we drink, and where we eat can be subsumed into the two dimensions, which Bourdieu called economic and cultural capital. Then the individuals can also be placed onto the graph as a 'cloud of points' too. However what Bourdieu did was rather different: he plotted 'variables' onto this graph (see Fig. 4.6). First, we can plot any variable that is part of the analysis, with its central mass point placed optimally to cover all the cases. Second, we can also plot any ordinal or binary supplementary variable onto this space too. The last plot is achieved by estimating for every case in each sub-group of the variable what the average index value is for factor 1 and for factor 2, and this is where the point is placed.

Regression and MCA Can Be Grounded on Weak Structuralism

▶   **Tip** Causality is a delicate subject because many causal claims can be attacked. If structural determination is only weak and the mechanisms you list do not dictate outcomes, your regression findings could be subject to criticisms. Mixed-methods research accepts that criticisms may occur and evidence is made available to help justify and defend the positions taken.

Multiple correspondence analysis uses 'clouds' on the one hand, and regression uses dependent and independent variable correlations, on the other hand. Regression and MCA are rather different. Both methods can be used to illustrate retroduction. In MCA, the researcher keeps iterating from questionnaire to bar charts, then to MCA, then back to the questionnaire, carrying out closed retroduction to decide on which variables to include. Meanwhile they also engage in open retroduction to make enquiries about constraint, paradox, contradictions, and cultural forms which are or sometimes are not matched by actual individual behaviours. (There are often exceptions to the broad cultural patterns.)

Regression too can be linked with interview data. This is easiest if we can interview a few of the same respondents who are in the primary survey. Often we don't have the luxury of matching them up, so we have to triangulate at the level of ideas, variables, causes, and causal associations, rather than by matching individual quotes to individual manifest variable values. If you do get a chance to run a small primary survey, be sure to do some semi-structured interviews and match up your data using NVIVO or a spreadsheet if you have time. The materials may turn out to be suitable for both statistical methods (regression and MCA) if you have suitable research questions that fit with these methods.

In conclusion, retroduction is used a lot and occurs very naturally as a follow-up to one's initial enquiries. So far, retroduction has not been widely treated in methods 'textbooks', being seen more as a topic for advanced philosophy of science. Yet no critique has yet coherently attacked this concept, either by scientists or by philosophers. Retroduction will soon get included in social research encyclopaedias. Retroduction appears in a growing number of influential books, for example Lawson (2003), Bridges et al. (2011), Sayer (1984, 1992, 2000), Byrne (2002), and others. Using a mixture of logics, including retroduction, you can feel firmly grounded in your knowledge. It could be expensive to use SMMR though, if you need to include both questionnaires (for doing MCA or regression) and some other methods such as interviews. For this reason, in later chapters I discuss the generality of the conclusions we can validly draw from systematic mixed-methods research. It may be possible to save on research costs by using secondary corpuses (Chap. 8) or by collapsing key data into tables for closer analysis (Chap. 6).

Review Questions

Give three examples where a social mechanism that causes an outcome to arise, or emerge, could lead to a specific variable to be recorded in an administrative survey or a questionnaire.

How is regression done? Explain it as a series of steps.

Write a paragraph supporting pluralist models in regression, and noting how mixed methods could contribute to knowledge about the cases, the variables, or the underlying causal mechanisms.

If we want to use cluster analysis or multiple correspondence analysis (MCA), we could provide variables about the underlying context of a behaviour along with measures of the behaviour itself. Make a list of six variables that all represent one group of behaviours and consider whether this list would work in an MCA. State whether or not you also need regression and explain why.

Further Reading

Greene (2007) helps researchers use mixed methods. Books on epistemology range from Bryman (1996), about balancing methods, to Outhwaite (1987), which is about qualitative interpretation. Kent (2007) explains research design for business topics using mixed methods. Dow (2004) and a series of papers by Downward and Mearman (2002, 2004, 2007) explain why triangulation is likely to offer a mixture of corroboration and discovery. Flick (1992) explains that validation is not the key purpose of triangulation. Retroduction was presented by Bhaskar (1979), Blaikie and Priest (2013), Layder (1993, 1998), Sayer (1992, 2000), and Olsen and Morgan (2005). Byrne (2005) discusses how configurations can be defined to make case-study research work well. Creswell and Plano-Clark (2007) helpfully cover some options for mixed-methods research.

Appendix

Table 4.3 Variates and variables in the exemplars

Variate and **source**	Abbreviation	Type	Status in article	Source
Lam and Ostrom (2010)				
Farmers developed fines for water misuse	F	Event	Independent variate	Lam and Ostrom (2010)
Whether a system received infrastructure assistance	I	Event	Independent variate	Lam and Ostrom (2010)
Farmers developed rules for system	R	Event	Independent variate	Lam and Ostrom (2010)
Whether leadership in a system has been able to maintain continuity and to adapt to a changing environment	L	Institutional	Independent variate	Lam and Ostrom (2010)
Represents whether farmers have been able to maintain a certain level of collective action in system maintenance	C	Institutional	Independent variate	Lam and Ostrom (2010)
Whether the watershed has a good water supply	W	Structural	Dependent outcome	Lam and Ostrom (2010)
Cress and Snow (2000)				
Viable social movement organisation (SMO)	Viable	Institutional	Independent variate	**Cress and Snow (2000)**
SMO used disruptive tactics	Disrupt	Event	Independent variate	
SMO had allies	Allies	Structural	Independent variate	
SMO had city support	City	Event	Independent variate	
SMO developed a diagnosis of homelessness	Diagnosis	Event	Independent variate	
SMO developed a prognosis for homelessness	Prognosis	Event	Independent variate	

(continued)

Table 4.3 (continued)

Variate and **source**	Abbreviation	Type	Status in article	Source
Formal representation of SMO members on decision-making board(s)	Representation	Member	Dependent outcome	
Obtained control over resources for SMO	Resources	Event	Dependent outcome	
Secured new rights for beneficiaries	Rights	Events	Dependent outcome	
Obtained facilities or restorative programmes	Relief	Events	Dependent outcome	
Byrne and Ragin (2009)				
And Byrne (2009)				
Catholic School	C	Institutional	Independent variate	
Socially deprived area	D	Structural	Independent variate	
Selective school admissions policy	SEL	Event	Independent variate	
High level of special needs intake	SEN	Event	Independent variate	
Good exam results	High	Event	Dependent outcome	
Rosigno & Hodson (2004)				
Conflictual coworker relations		Institutional		
Workforce over 50% female in this unit	W	Structural	Independent variate	
Workforce over 50% from minority ethnic group	M	Structural and institutional		
Bureaucratic: procedural rules are written	B	Institutional	Independent variate	
High worker input into task organisation	A	Institutional	Independent variate	
Frequent verbal or other abuse by supervisors	Abuse	Event	Independent variate	
Fulfilling work = sense of purpose and meaning in work		Event	Dependent outcome	
Job satisfaction		Event	Dependent outcome	
Pride in work		Event	Dependent outcome	
Commitment to organisational goals		Event	Dependent outcome	
Procedural sabotage		Event	Dependent outcome	
Infighting within work groups		Event	Dependent outcome	
History of workers striking		Event	Dependent outcome	

	Abbreviation	Type	Status in article	Source
Kanjilal-Bhaduri and Pastore (2018)				
Age		Structural	Independent variate	Kanjilal-Bhaduri and Pastore (2018)
Dependent children		Institutional	Independent variate	Kanjilal-Bhaduri and Pastore (2018)
Dep. children interacted with gender		Event	Independent variate	Kanjilal-Bhaduri and Pastore (2018)

(continued)

Table 4.3 (continued)

	Abbreviation	Type	Status in article	Source
Gender		Event	Independent variate	Kanjilal-Bhaduri and Pastore (2018)
Labour market work experience		Event	Independent variate	Kanjilal-Bhaduri and Pastore (2018)
Education		Institutional	Independent variate	Kanjilal-Bhaduri and Pastore (2018)
Ethnic minority		Institutional	Independent variate	Kanjilal-Bhaduri and Pastore (2018)
Whether they do paid work		Event	Dependent variable	Kanjilal-Bhaduri and Pastore (2018)
Wage rate (adjusted, per hour)		Event	Dependent variable	Kanjilal-Bhaduri and Pastore (2018)
Variables not shown here		Mixture	Independent variate	Kanjilal-Bhaduri and Pastore (2018)
Training		Institutional	Independent variate	Kanjilal-Bhaduri and Pastore (2018)
Widow(er) status		Structural	Independent variate	Kanjilal-Bhaduri and Pastore (2018)
Marital status		Structural	Independent variate	Kanjilal-Bhaduri and Pastore (2018)
Household per capita expenditure		Structural	Independent variate	Kanjilal-Bhaduri and Pastore (2018)

See also Olsen and Mehta 2006a and b.

	Abbreviation	Type	Status in article	Source
Kelly et al. 2011				
ROWE	Membership	Treatment	Kelly et al. (2011)	
Gender interacted with ROWE	G-ROWE	Condition+event	Treatment mixed with structural independent variable	Kelly et al. (2011)
GADC	Condition+event	Independent variable mixture		
ROWE interacted with GADC	ROWE-GADC	Condition+event	Treatment mixed with structural independent variable	Kelly et al. (2011)
Gender of worker	Gender	Structural	Independent variable	"
Gender interacted with ROWE	G-ROWE	Condition+event	Treatment mixed with structural independent variable	"
Age	Age	Structural	Independent variable	Kelly et al. (2011)
Income	Wage	Structural	Independent variable mixture	Kelly et al. (2011)
Length of job tenure	Tenure	Event	Independent variable mixture	Kelly et al. (2011)

(continued)

Table 4.3 (continued)

	Abbreviation	Type	Status in article	Source
Managerial job	Manag	Institutional	Independent variable mixture	Kelly et al. (2011)
Total hours	Hours	Event	Independent variable mixture	Kelly et al. (2011)
Schedule control at baseline date	S1	Condition	Independent variable mixture	Kelly et al. (2011)
Schedule control after ROWE sessions	S2	Event	Outcome variable	Kelly et al. (2011)
Work-family conflicts	Opposite of work-life family balance	Event	Outcome variable	Kelly et al. (2011)
Hala Abou-Ali et al. 2010				
Mixed social fund treatment		Event	Treatment	
Other interventions		Event	2nd simultaneous treatment	
Access to sewage system (2004/2005)		Structural	Independent variable	
Female-headed household (%, 2004/2005)		Institutional	Independent variable	
Illiteracy rate in household (%, 2004/2005)		Institutional	Independent variable	
Household size (average, 2004/2005)		Structural	Independent variable	
Local training prevalence (2004/2005)		Institutional	Independent variable	
Poverty rate in 2000		Structural	Independent variable	
Poverty headcount rate in 2004/2005		Event	Outcome variable	

Note: A variate is a concretely measured feature of a case; a variable is a measure which exists both for the cases that were in a survey and for further potential cases, either actual cases that exist or potential ones and future ones, known as the super-population.

References

Bayliss, David, Wendy Olsen, and Pierre Walthery (2015) "Well-Being During Recession in the UK", *Applied Research in Quality of Life*. Online first: 29 April 2016 http://link.springer.com/article/10.1007/s11482-016-9465-8.

Bhaskar, Roy (1979) *The Possibility of Naturalism: A Philosophical Critique of the Contemporary Human Sciences*. Brighton: Harvester Press.

Bhaskar, Roy (1997) *A Realist Theory of Science* (2nd ed.). London: Verso.

Blaikie, Norman, and Jan Priest (2013) *Designing Social Research. 3rd Edition*, Cambridge: Polity (2nd ed. 2009)

Bourdieu, Pierre (1979) *Distinction: A Social Critique of the Judgment of Taste*, Translated from French by Richard Nice, 1984; 1st ed. in Classics series 2010, London: Routledge.

Bourdieu, Pierre (1995) translated from French 2005, *The Social Structures of the Economy*, Cambridge UK: Polity Press.

Bridges, Sarah, David Lawson, and Sharifa Begum (2011) "Labour Market Outcomes in Bangladesh: The Role of Poverty and Gender Norms", *European Journal of Development Research*, 23:3, 459–487.

Bryman, Alan (1996, original 1988) *Quantity and Quality in Social Research*, London: Routledge.

Byrne, D. (2002) *Interpreting Quantitative Data*. London: Sage.

Byrne, D. (2005) "Complexity, Configuration and Cases", *Theory, Culture and Society* 22(10): 95–111.

Byrne, D. (2009) "Using Cluster Analysis, Qualitative Comparative Analysis and NVivo in Relation to the Establishment of Causal Configurations with Pre-existing Large N Datasets: Machining Hermeneutics". Pp. 260–68 in *The Sage Handbook of Case-Based Methods*, edited by D. Byrne and C. Ragin, London: Sage Publications.

Byrne, David, and Ragin, Charles C., eds. (2009) *The Sage Handbook of Case-Based Methods*, London: Sage Publications.

Cress, D., and D. Snow (2000) "The Outcome of Homeless Mobilization: the Influence of Organization, Disruption, Political Mediation, and Framing." *American Journal of Sociology* 105(4): 1063–1104. URL: http://www.jstor.org/stable/3003888

Creswell, J. W. and V. L. Plano Clark (2007) *Designing and Conducting Mixed Methods Research*. Thousand Oaks, California and London: Sage.

Danermark, Berth, Mats Ekstrom, Liselotte Jakobsen, and Jan Ch. Karlsson, (2002; 1st published 1997 in Swedish language) *Explaining Society: Critical Realism in the Social Sciences*, London: Routledge.

Dow, S. (2004) Structured Pluralism, *Journal of Economic Methodology*, 11: 3.

Downward, P. and A. Mearman (2002) "Critical Realism and Econometrics: Constructive Dialogue with Post Keynesian Economics." *Metroeconomica* 53(4): 391–415.

Downward, P. and A. Mearman (2004) "On tourism and hospitality management research: A critical realist proposal." *Tourism and Hospitality Planning Development* 1(2): 107–122.

Downward, P. and A. Mearman (2007) "Retroduction as mixed-methods triangulation in economic research: reorienting economics into social science." *Camb. J. Econ.* 31(1): 77–99.

Flick, Uwe (1992) "Triangulation Revisited: Strategy of Validation or Alternative?" *Journal for the Theory of Social Behaviour* 22(2): 169–197.

Flyvbjerg, Bent (2011) *Making Social Science Matter: Why Social Inquiry Fails and How it Can Succeed Again*, Cambridge, UK: Cambridge University Press.

Greenacre, Michael, and Jorg Blasius, editors (2006) *Multiple Correspondence Analysis and Related Methods* (Series: Chapman & Hall/CRC Statistics in the Social and Behavioral Sciences), London: Chapman and Hall/CRC.

Greene, Jennifer C. (2007) *Mixed Methods in Social Inquiry*, San Francisco: John Wiley & Sons (via Jossey-Bass Publishers).

Abou-Ali, Hala, Hesham El-Azony, Heba El-Laithy, Jonathan Haughton & Shahid Khandker (2010) "Evaluating the impact of Egyptian Social Fund for Development programmes", *Journal of Development Effectiveness*, 2:4, 521–555, http://dx.doi.org/https://doi.org/10.1080/19439342.2010.529926

Kanjilal-Bhaduri, Sanghamitra, and Francesca Pastore (2018), Returns to Education and Female Participation Nexus: Evidence from India, *The Indian Journal of Labour Economics* (2018) 61:515–536. https://doi.org/10.1007/s41027-018-0143-2.

Kelly, Erin L., Phyllis Moen and Eric Tranby (2011) Changing Workplaces to Reduce Work-Family Conflict: Schedule Control in a White-Collar Organization, *American Sociological Review*, 76:2, 265–290: https://doi.org/10.1177/0003122411400056.

Kent, Raymond, Chapter 15 of R. Kent (2007), *Marketing Research: Approaches, Methods and Applications in Europe*, London: Thomson Learning.

Lam, W. Wai Fung and Elinor Ostrom (2010), "Analyzing the dynamic complexity of development interventions: lessons from an irrigation experiment in Nepal, *Policy Science*, 43:1, pp. 1–25. https://doi.org/10.1007/s11077-009-9082-6 .

Lawson, T. (2003) *Reorienting Economics,* London and New York: Routledge.

Layder, Derek (1993) *New Strategies in Social Research,* Cambridge: Blackwell Publishers. Also Repr. 1995, 1996, Oxford: Polity Press.

Layder, Derek (1998) "The Reality of Social Domains: Implications for Theory and Method", ch. 6 in May and Williams, eds. (1998), pp. 86–102.

McNiff, Jean, and A. Whitehead (2011) *All You Need to Know About Action Research,* 2nd ed., London: Sage.

Mukherjee, Dipa (2012) Schooling, Child Labor, and Reserve Army: Evidences from India, *Journal of Developing Societies*, Vol 28(1): 1–29. https://doi.org/10.1177/0169796X1102800101.

Olsen, Wendy (2012) *Data Collection: Key Trends and Methods in Social Research*, London: Sage.

Olsen, Wendy (2019) "Bridging to Action Requires Mixed Methods, Not Only Randomised Control Trials", *European Journal of Development Research*, 31:2, 139–162, https://link.springer.com/article/10.1057/s41287-019-00201-x.

Olsen, Wendy, and Smita Mehta (2006a) "The Right to Work and Differentiation in Indian Employment", *Indian Journal of Labour Economics,* 49:3, July-Sept., 2006, pages 389–406.

Olsen, Wendy, and Smita Mehta (2006b) A pluralist account of labour participation in India. Online mimeo, GPRG-WPS-042. URL http://www.gprg.org/pubs/workingpapers/pdfs/gprg-wps-042.pdf, accessed Jan 2019.

Outhwaite, William (1987) *New Philosophies of Social Science: Realism, Hermeneutics and Critical Theory*, London: Macmillan.

Porter, Sam (2011), Chapter on "Realism", in L. Miller & John Brewer eds. *The A-Z of Social Research* (pp. 256–259), London: Sage Publications. https://doi.org/10.4135/9780857020024.

Rabe-Hesketh, Sophia, and Brian Everitt (2009) *A Handbook of Statistical Analyses Using STATA,* London: Chapman and Hall/CRC.

Rosigno, V.J., and Randy Hodson (2004) The Organizational and Social Foundations of Worker Resistance, *American Sociological Review* 69(1):14–39 https://doi.org/10.1177/000312240406900103.

Savage, Silva, Warde and Gayo-Cal and Wright (2009) *Culture, Class, Distinction*, London: Routledge.

Sayer, Andrew (1992 (orig. 1984)) *Method in Social Science: A Realist Approach*. London, Routledge.

Sayer, Andrew (2000) *Realism and Social Science*. London: Sage.

Silva, Warde and Wright (2007) Using Mixed Methods for Analysing Culture: The Cultural Capital and Social Exclusion Project, *Cultural Sociology*, 3:2, July, pages 299–316, 10.1177/1749975509105536.

Factor Analysis in a Mixed-Methods Context

5

This chapter takes up factor analysis which reduces a very wide set of variables into a single 'latent variable' such as an index of intelligence. I will call the latent variable an index scale, since it is a continuous variable. In this chapter, I explain the basic logic of factor analysis and explore its ontological underpinnings. The two main approaches historically were latent factor analysis (LFA) and principal components analysis (PCA). In recent years a further range of possibilities opens up by allowing for ordinal measured variables, which are reduced using confirmatory methods, so I will provide an up-to-date summary of the possibilities. 'Structural Equation Modelling' (SEM) is the quantitative method that helps us make a path diagram and work out direct and indirect associations between independent and dependent variables. The structural equation model can be viewed as a highly flexible umbrella system for regression methods. Qualitative methods could be used at several stages in the LFA protocol, as I will show. The triangulation chosen will influence whether the project is predominantly exploratory, retroductive, or confirmatory latent factor analysis.

'Retroduction' here is potentially controversial, as the specific retroductive logic affects how we create and interpret latent factors. Retroduction will be crucial for making discoveries and having holistic knowledge using latent variable analysis. This chapter provides a sample of research design plans for different situations.

5.1 Latent Variables and Entities

▶ **Tip** Moving towards the use of a latent variable is a thought exercise. Do not be afraid to be innovative. On the other hand, the available variables may not hold up well in latent variable analysis. Check the manifest variables carefully before you proceed.

Latent variable analysis—which seeks latent variables from a set of variables—allows the creation of a scale that summarises a large table of numbers. The

© The Author(s), under exclusive license to Springer Nature Switzerland AG 2022
W. Olsen, *Systematic Mixed-Methods Research for Social Scientists*,
https://doi.org/10.1007/978-3-030-93148-3_5

resulting scale can be called a latent variable because it wasn't measured directly when the other variables were recorded. One could illustrate latent variables by imagining first a summary of the taste and price features of beers. If each beer type is the 'entity', the characteristics of beers are suited to measurement using a latent variable. For example, the latent variable could be a scale that is high for a good-tasting, expensive beer; and the latent variable could take a low value for awful tasting cheap beers. The character of the beer is part of its essence.

In clarifying what we mean by 'a beer' we are doing 'casing'. Casing means defining what key elements together constitute one case, here for instance a liquid, sold as beer, produced by a particular company under a specific brand name. Examples would be Coors, Budweiser, and Holts beers. The sense in which the taste of a beer is its essence is partly that if a beer failed its taste test or was impure, the company would hold it back. The companies only release beers under a brand name if they hold the ingredients that are listed, and meet the standards set, for that particular beer. See Ragin (2009) which explains how casing is done.

Another way to represent a scale using a latent variable is to offer a summary of student scores on a psychometric test of 25 items. The average person would have a latent score near the average. A few people would have a very high score, and a few people would have a very low score. A third example is making a personal leadership qualities scale. Here we need items which cover people's leadership traits and leadership activities. The resulting single score, one per person, summarises the whole questionnaire segment on this topic. In other words, latent variables carry out data reduction to a more manageable single variable. It is a new variable and it will be called a latent, derived, or predicted variable. The latent variable score for each person is a single number.

In terms of the data matrix, for survey data the new latent variable is a new column. It is 'saved' by the statistical package. It is created from three or more other columns in the same data set. Typically the number of cases (N) in the data is the same as the number of scale index values predicted.

Latent factor analysis, also called confirmatory factor analysis (CFA), will be used when you need to reduce your data set down to a smaller number of variables. Figure 5.1 shows how three variables might reduce to one latent variable. It is also useful if, on the other hand, you simply wish to get an index scale that summarises some aspect of the cases in the data.

Thinking of data reduction, an example is that if you have attitude data on ten attitudes, most of which are indicative of one broad area such as 'egalitarian gender attitudes', you can make a latent variable. The word 'latent variable' refers to the result, which is analogous to a scale or index, running either from -3 to $+3$, or from 0 to 100. The latent variable can be set out onto a wider scale to match some other units, such as currency, after it is created. The aims of the factor analysis may be varied or even mixed, while the mathematics is common as shown in Fig. 5.1. In this figure we imply that a regression is carried out by adding arrowheads. The regression however is non-standard, because we have three equations (as shown), rather than one. Each manifest variable is regressed upon the latent factor. (For easy reading we will call the input variables the 'manifest' variables, since they were actually

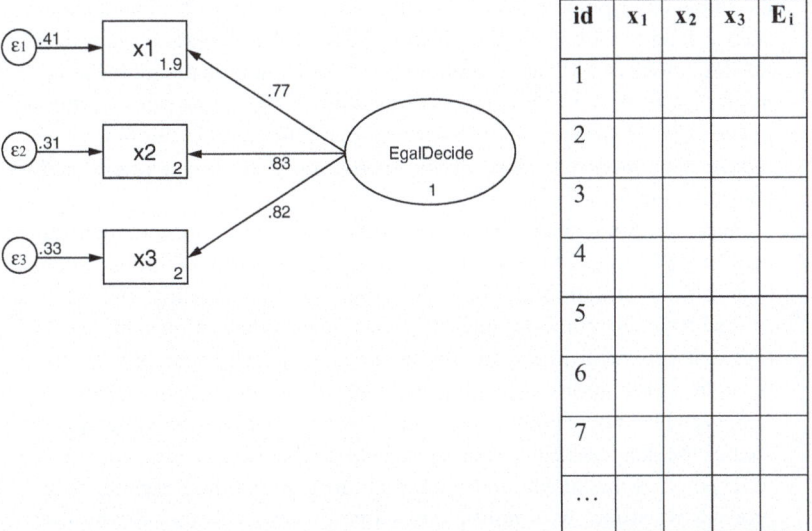

id	X₁	X₂	X₃	Eᵢ
1				
2				
3				
4				
5				
6				
7				
...				

Fig. 5.1 A latent variable and its manifest variables. Note: This latent variable is shown with three continuous manifest variables. Because we do factor analysis to discover the latent variables, I sometimes call these latent factors. The latent variable is labelled E for 'egalitarian decision-making attitude' in the table. E is a derived variable not a measured survey variable. Individuals numbered 1–8 in the identification variable have their own attitudes. The social norm around this matter is measured by the mode of the resulting latent variable E. Deviance from the norm is possible for individual attitudes, which are factor scores E_i. A wide variety of software choices exist for factor analysis (some software packages are SPSS, Latent Gold, MPLUS, STATA or R; see Field, 2013 for the use of SPSS, and Hamilton, 2006 for Stata). Below, I summarise five kinds of latent variable.

measured.) The computer is able to arrive at the ideal factor scores by weighting the manifest variable values for each case, and thus obtain the best fitting latent variable.

Five Kinds of Latent Variable

▶   Type 1: The latent variable measures a social norm around its median. The level of the variable for each unit, such as a person, measures personal attitudes which may deviate from the norm. Instead of being measurement error, the distances from a measure of F for person 1 or 2 to the median of F are real deviations. Deviance of attitudes from norms will tend to be unusual, so we may get a unimodal latent variable. On the other hand, however, some norms have deviation only in one direction, leading to a skewed distribution. The norm at one end (how well do you agree with war being bad?) is distinguished from the wide range of the relatively rare responses (war is good—the respondent has rationalised war and, as a result, reports a personal attitude that deviates from the norm).

Type 2: In general, latent variables measure traits of entities. Again the mean, mode, median, and skewness may represent real features of the range of responses. Entities such as airplanes have correlated traits such as length, width, weight, maximum speed. The general issue is that the deviations of individual units, such as the Concorde plane, from the norm may be real; or there could instead be measurement error leading to variance. Measurement error on each manifest variable might tend to reduce the observed correlations—particularly if it is uncorrelated measurement error.

Type 3: Spuriously associated indicators of a scene, such as measuring features of house fires or crime scenes or other entities. It can be a difficult situation in which the 'traits' are conveniently measured and do have correlations, suggesting one underlying entity with a key trait that gets measured using the latent variable. The difficulty would be if the correlations were due to confounding unmeasured features (e.g. variables not available might have covered the broader geographic area or the dangers of the appliances held by the owner of the house).

Type 4: A latent variable could group together key aspects of a process over time, with multiple possible processes discovered by the pattern of co-correlation. For example in treating Accident and Emergency cases, which fall into a typology, we could discover the typology by inputting as manifest variables all features of all cases. Groups of 5–6 features might indicate each main type of accident case: bone injuries, cardiac emergencies, and so on. The analysis of the data could help to classify the units (cases) into the best typology by levels of the variables for each case. Discriminant analysis or multi-trait multi-class analysis could also be applied in such situations. Measurement error or real deviance from each of the main types in the typology might apply, and therefore some confusion can arise over poor fit situations. Notably it is possible for one case in a hospital setting to present with multiple overlapping conditions.

Type 5: Another medical application is when socially agreed sets of symptoms are mapped onto key disease names to create a classification system. These systems of naming diseases via their symptoms are highly politicised and raise contestation and debate. Some medics may prefer to name a disease according to its origins or explanatory factors, or according to genetic patterns. Therefore, medical disease names have become a historical genre of their own.

All the above types of latent variable raise issues of the level of measurement of the manifest variables, which may be binary, ordinal, or continuous, and can be very numerous.

Entity realism helps us with all the above types of latent variable. We usually embed the latent variable estimation into a wider process, shown in more detail in Fig. 5.2. In the first column a simple latent variable model is offered. The middle column shows how useful the PCA method is when you are at risk of being overwhelmed by data. If there are 35 columns of data, the project may be intractable.

PCA helps simplify the data down into its main conceptual areas. Each gets one index. I have added notes that might correspond to either confirmatory or exploratory latent variable analysis. You can test for group-wise differences in means or variances within your sample. These methods offer rich opportunities. It is also possible to test a local primary survey sample against a baseline population estimate. This will require two CFA runs. First run it on the national data. Secondly, after administering the same, or similar, survey on your local cases, run the latent variable analysis on them too. You may compare either the factor scores, factor loadings, or means of groups.

In Fig. 5.2, double-headed arrows can indicate correlations (curved lines) and/or bi-directional relationships (straight lines).

You can use a latent variable for egalitarian gender attitudes to summarise all ten or perhaps just seven to eight of the available variables. Figures 5.1 and 5.3 illustrate the latent variable that arose as an estimate, bringing together four items from a

Confirmatory Factor Model:	Exploratory Factor Model:	Example of Closed Retroduction:
Assert that N manifest variables X1-X3 are indicators which are 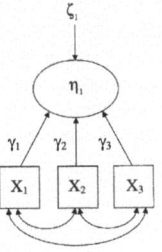 associated with higher or lower levels of the underlying factor, η (eta). Measure the factor loadings γ1 to γN. Measure goodness of fit. Allow for correlations of the manifest variables. In CFA you choose how many factors to fit (and just 1 factor is shown).	Assert that N factors may exist in society, and that the measures X1 to X35 might correspond to these. N is an unknown. Explore the correlations of X1 to X35, grouping those with strong correlations. For binary variables, use an alternative measure, φ (Phi). For ordinal with ordinal, use Cramer's V. Decide on the best levels of measurement, then discover the optimal number of factors. You might conclude with 2 factors and ignore the rest of them.	Develop a factor analysis, then explore the possible alternative variables or alternative levels of measurement of the manifest variables. You may arrive at a better fit. Next you might also consider reasons for the factor being higher in one group or another (Group Tests). Two tests take place. 1) if groups differ in their factor loadings, create the factor independently in each group. 2) and/or if groups have the same factor loadings, they may have different means, so test for that.
The result is based mainly on the variation in the manifest variables. Measurement error is dropped out of the model.	The result is based on all the variation (common and unique) in all the manifest variables.	The result will cover a broader spectrum, such as regression, group tests, or a structural model.

Fig. 5.2 A confirmatory factor analysis (CFA) model and variants. Notes: Single-headed arrows reflect a measurable correlation. The dual-headed arrows show where the residuals of the regression equations are correlated. Lastly one residual is shown at the top (the latent variable residual error)

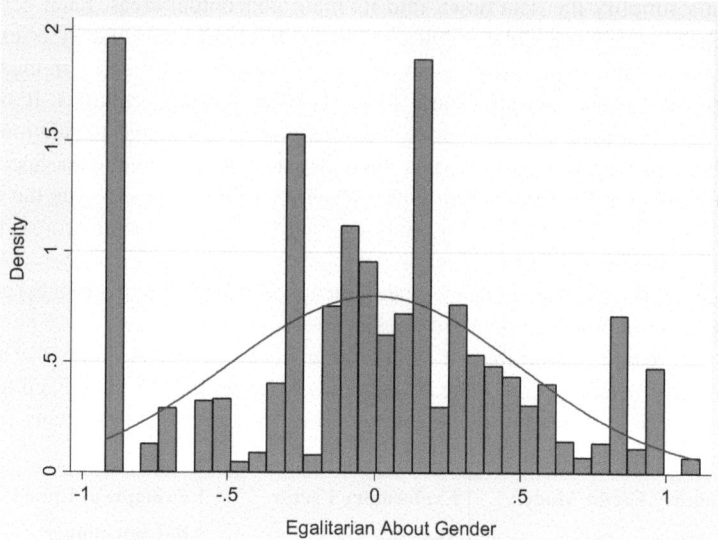

Fig. 5.3 A latent variable result based on four ordinal manifest variables (India). Sample coverage: Women aged 15 to 49. Key: The egalitarian attitudes measure is derived from four questions, each having five unranked answers, which we ranked according to whether they reflected male power, equal power of husband and wife, or female power. This ranking involved recoding the results so that 'male makes most decisions' was marked as least egalitarian and involved highest male power; joint decision making was in the middle; and female decision making was marked as most egalitarian and involved lowest male power. Full details in Olsen and Zhang 2015). (Source: Olsen and Zhang 2015, using data from the Demographic and Health Survey for India, known as National Family and Health Survey (full details are in ibid.)

questionnaire administered in India and Bangladesh. The analysis of covariance and co-associations helps you discern which variable(s) is (are) best to include in the latent variable. The defaults set in most software packages will tend to cause the variance of the scale to be 1 and the mean of the scale to be set at zero.

Figure 5.3 illustrates another latent variable.

On the other hand, sometimes when you want to get an index scale to bring together a set of variables, the research project might focus on the variance of this scale. This variance could be narrower than the variance of each of the inputted variables. It is also true that the variance might be lower in one social group than another: for example one religious group may have strongly normed attitudes, with a small variance, and all remaining groups may have a wide variance due to more diversity in attitudes. An example of a religious norm might be the degree to which someone believes it is crucial to pray or attend services or worship regularly; this norm might be called 'religiosity'. This norm is not the same as the actual frequency of worshipping. Attitudes will vary around the median of this norm. At the same time, it is possible to also measure religious activity; that might lead to constructing a second latent variable. (One could use frequency of attendance, frequency of study, number of religious books, and spending per year on religious activities.)

As shown in Fig. 5.1, the variance of a latent variable would be a measure of the horizontal dispersion of the index scale. In Fig. 5.1, the large clump of cases with a low value would cause this latent variable to have a high variance. In fact the traditions of the South Asian countries covered here have created the apparently trimodal shape in the egalitarian norm variable: three high lumps where cases are more common at a certain index value. Details are found in an online briefing paper (Olsen and Zhang 2015).

Each latent variable tends to bring together the common variance. The latent variables tend to omit the variation that is unique to just one of the variables. Initially this variation appears as a residual for that variable. The equations that underpin a latent variable analysis also allow 'residual' error for each latent variable. (Both CFA and PCA can work on situations of multiple latent variables. If you plan only one, then the distinction does not apply, and you are doing CFA.) A question arises whether the errors of the latent variables will be correlated. There are minor differences in how variability—that is, the dispersion—of the variables is treated in the two approaches, CFA and PCA. In using confirmatory factor analysis, the cross-correlations of the residual error of multiple latent variables are assumed to be zero, or else they have to be explicitly modelled and estimated (Muthén 1983; Muthén and Kaplan 1985; Muthén and Muthén 2006). In 'principal components analysis' (PCA), these correlations can run any direction, and they are not estimated explicitly. The software routine in PCA aims to get rid of these correlations by adding more unique latent variables. It is often said that in confirmatory factor analysis the new latent variables are uncorrelated with each other, but this is a requirement that can be relaxed. The PCA methods tend to reproduce the covariances of all the observed manifest variables, while assuming that the latent variables have no positive or negative correlations with each other.

The cases in a dataset need to be comparable before creating a latent variable. If you have people in the data, you might develop a personality-trait scale. If you have businesses, as 'cases', you can develop a scale to reflect profitability, growth patterns, or export-orientedness. The latent variable often plays the role of an adjective: it can be used to describe features of the underlying entities. The entities need a certain amount of homogeneity, and are usually referred to by nouns: households, cases, persons.

When we refer to entities like this, we make a dual assumption that they are comparable in certain key respects, and that gathering data on them makes sense. We do not usually mix people with businesses in a sampling list for a questionnaire. Instead we compare like with like. At the same time, as a duality, we are assuming there is enough heterogeneity among the cases for it to make sense to ask, record, or measure their response to some prompts. It is usually a situation where it is difficult to carry out these measurements that generate the need for factor analysis. For example, we cannot expect to get valid answers if we ask directly, 'How strong are your leadership traits?' Instead we ask more indirect questions. This gets the respondent to reveal their underlying traits.

The latent variable approach will seem complex to readers. A simpler approach is the 'classical scale': you just add up variables to get a score directly. An example will help show which is useful in selected contexts.

If you have measures of all the values of each and all household appliances owned by 1000 households, you may create a scale of appliance values directly. You can either add up how many appliances they have (i.e. a count variable), or you can add up the values (i.e. measure them all in currency units such as $). On the other hand if the measures of assets are very diverse and are mixed in their level of measurement, such as the four listed below, then you may decide to choose factor analysis.

Four asset measures:

- How many cars they have.
- How large their house is in square meters.
- How many mobile phones the household members have.
- Whether their house has plumbed running water.

These assets are commonly used in making a wealth index that allows social classes to be compared.

5.2   One Could Use Exploratory or Confirmatory Factor Analysis

▶   **Tip** The notion of 'realism' is useful with latent variables, because a realist argues that the latent factor pre-exists the measurement step. Therefore replication will tend to reproduce similar results for a factor analysis. Realists tend towards confirmatory, theory-informed variants of factor analysis.

In measuring the factor, the two traditional approaches to this problem are 'principal components analysis' (PCA) and confirmatory latent factor analysis (CFA). If you want to use PCA, you would combine the measures listed above and let the computer work out what 'weight' to give to each variable. One wants to have clarity about the unit of analysis. The unit of analysis here is the household, not the person. (And it is not a business.) With this clarity, we can allow the sum of mobile phone counts to appear here. PCA can be used for 4, 8, 36, and even larger numbers of manifest variables. It will require you to allow >1 factor to arise. Some users advise that you then use the 'first factor' as your key scale. For 7 variables, the PCA may have both a first and second factor, and these will (perhaps) exhaust the variation in the data.

Confirmatory latent factor analysis (CFA) works differently, and it requires that you decide how many latent variables to have. The CFA will seek the set number of latent variables, often 1, 2, or 3, and then let remaining variation be unexplained variation. These 1 to 3 latent variables are the ones you have already theorised. An advantage of CFA is that it brings to the centre of attention that each manifest variable has an error term. Measurement error is allowed for.

On the other hand, PCA is usually not well theorised. It is mainly an algebraic method aimed at reducing a set of variables to a much smaller set of new variables. But CFA is more theorised from the start, and it invokes social theory explicitly. CFA is commonly used in social science settings. This theorising is not causal theorising. It is about what correlates with what; what measures reflect or respond to underlying conditions; what is similar to what. A well-theorised CFA will have high correlations of the manifest variables. A PCA may have some low correlations, which create distinct latent variables in a multi-factor result. One can also try to have a multiple factor CFA. Here the theory would assert for example 'beers are cheap or expensive, based on multiple price points across the regions' and 'beers taste good or bad according to local tastes, across the regions'. In this example two factors would arise. They may be poorly correlated with each other, which is fine in both CFA and PCA. Then we can find the cheapest good beer, since each beer has 2 factor scores.

Suppose you want to measure well-being. Your reading suggests this will include aspects of satisfaction with one's life generally, feeling well in oneself and being free from mental strain. We may say these three manifest variables include two positive and one negative indicator for well-being. The negative one would be a question like 'Have you felt under strain in the last six months?' One seeks out the three questions in some data, perhaps a bespoke survey or a large-scale panel survey dataset.

I would like to comment on two facets of latent variable analysis that have arisen so far. First, the entities are not the actual latent variables, nor are they the index scales, but rather they are underlying things such as social norms. Second I will comment on the idea of errors in measurement and what these (ontologically speaking) refer to. In both cases, I will be speaking about the reality as if the data corresponded to the reality, rather than speaking about the latent variable as the reality. A measurement is not the same as the thing being measured.

A famous painting from Belgium once had as text, written below the image of a pipe, *Ceci n'est pas un pipe*, or *This is not a pipe*. This was far from a joke. The painting became part of a long running debate in both art and philosophy: if the painting was merely a representation, then it did not have the capacities and liabilities of real pipes; but it had other capabilities, such as the ability to entertain or to stimulate the viewer's emotions or our creativity. At one and the same time this image and text raised two issues. Firstly the meanings of symbols and signals, which are very important in the sociological and anthropological sciences. The message was that the symbol has its own role, and is affecting you through your eyes, mediated by your socially shaped brain; and is at the same time an intended stimulus created by a human in their socially shaped location. At one and the same time we may have, in our mind's eye, real pipes of the past, with their smell and their hardness and price and owners. Therefore, whilst this is *not* a pipe, it is a symbol that corresponds to pipes and may in fact act as a reminder of pipes.

For some scholars there is a debate about whether pipes exist at all. Perhaps 'pipe' refers only to an ideal type or non-extant thing whose thingness is imaginary. The notional pipe may be too pure and refined to match any real, carved pipe of wood. On the other hand, perhaps 'pipe' refers to something really important and powerful: a piece of wood that bears smoke and can be influential on the user's health, the nearby air quality, and the smell organs. The 'power' to influence the surrounding area is known as a capacity or capability in realist language. Various entities such as pipes and paintings are seen as having capacities and being liable to certain outcomes by dint of their essential qualities. See Potter (2000), Lawson (1997, 2003), and Sayer (2000) for more discussion. We can use care and discernment in deciding what we can represent using measurements in either a direct way, or in the latent-variable manner. There are sure to be some aspects of society that do not fit measures or binary variables well at all. Conflict and debate, for example, may not suit algebraic representations. Borsboom et al. (2003) argued that latent variable analysis is good for measuring the traits of entities, yet it is not very good at measuring how processes work. In the rest of this chapter, I will use examples that suit latent variable analysis fairly well.

I can discern at least four kinds of things that latent variables represent. Whatever your discipline you can probably think of examples of each one. Based on the exemplars in Chap. 2 I present a few examples below.

- A social norm, with attitudes spread out around it both above and below the mean.
- A combination of diverse dissimilar objects which fall in a group or continuum on some essential aspect. Example of asset index.
- An aggregation of things whose scale value themselves could cumulate, but we optimise this scale so that it fits well but is not too skewed. Example of a wealth index, a latent variable based on a series of asset values but instead of summing them, takes a weighted average resting upon underlying correlations of high and low values.
- A measure of the trait(s) of an entity. Example of work life balance: given the hours worked, overtime paid and unpaid, and time spent on family duties there is a latent variable for WLB above which is overwork and below which is the luxury of relative underworking.
- Symptoms of a medical condition.

In all these cases we are measuring traits of entities. The abstraction step is important. Variations can be used, such as simply averaging out scores on four psychometric tests. Latent variable approaches are most useful if score averages have not already been measured.

5.3   Measurement Issues for the Manifest Variables in a Confirmatory Model

When we set up a research project to carry out CFA with local primary data, or national random-sample data, we need to be very clear about the broad arena of the research (Fig. 5.4), then carry out the enquiry and move to an interpretive stage. As

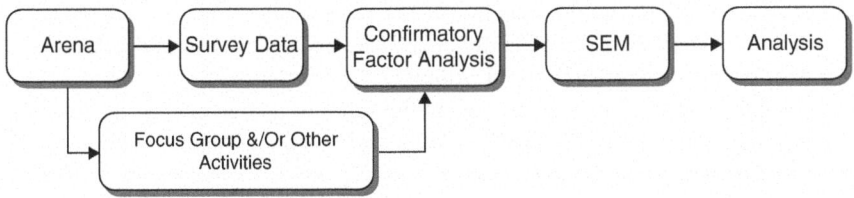

Fig. 5.4 Systematic mixed-methods research with confirmatory factor analysis. Notes: 1: The use of ordinal variables is arguably superior to requiring each variable to be continuously, cardinally measured. It is certainly more flexible and it is a situation that commonly arises. 2: Your underlying qualitative research is extremely useful for interpreting the factors. Key: SEM = Structural Equation Modelling

usual, the research design might move through these stages rapidly, or you may have time for feedback loops (Chap. 3). The 'arena' is my phrase for all the onto-logical fleshing-out of the research question(s), discerning the units of analysis, multilevel analysis if necessary, and for a survey you will also need to choose or write fresh questions and work out a coding plan.

See also in Chap. 3 the distinctness of the arena stage, data stage, and concluding stage of a mixed-methods project. When using CFA we make firm decisions about cases, levels, and measures at the arena stage. Based upon these, we measure the goodness of fit of a latent variable. The other activities such as interviews and focus groups might be used either before, during, or after the factor analysis, or all three. But 'confirmatory' approaches eventually settle down to declare the conceptual framework that underpins the choice of the variables.

5.4   Mixed-Methods Research Designs Using Latent Variables

Mixed methods with innovative research methods is important for enriching your latent factor analysis model.

In the past, some users of secondary data sets would explore the data—that is search through variables prior to theorising—looking for items that might fit into one index scale. This approach fits well with exploratory factor analysis. A simple example is when we want an index of wealth, but we do not have the value of each asset. We could instead make a big list of all the assets and then let the computer decide which ones belong in the first factor. (This would be principal components analysis, PCA, or exploratory factor analysis EFA.) In my view, such an approach is not advisable for modern scholars because it would be poorly theorised. Yet the use of EFA in general, and PCA in particular, is growing in a fresh wave of popularity due to the rise of large datasets under two remits:

- Data science administrative data in general. Data science often uses merged versions of existing data, as originally created by an organisation for its own purposes. We create a research result from it.
- Consumer research in particular, where the seller or a large buying organisation has a huge database of correlated measures of behaviour, and they do not have time or inclination to theorise about it.

In both cases it may make sense to use computers to 'reduce' the data and then interpret what emerges. Data reduction via exploratory factor analysis is not led by the purposes, theories, aims, or intentions of the researcher. Any factor could emerge from the group of manifest measures. The error-minimising approach is typically taken. I do not find this approach as useful as thinking carefully about operationalisation useful, meaningful concepts that refer to real entities.

For many social scientists the exploratory factor analysis is too much like a black-box approach. What goes in, comes out. It makes a mystery of the grouping of variables, and then we have to unravel this mystery.

Once the data and its underlying topic are understood very well, we will be able to pick out the relevant variables, group them and order them, and discern the things represented by the latent factors. Each latent variable is known within a theoretical construct, and only later is measured. We also develop an S I M E type of model so that the explanatory and the outcome variables are separated, so that we can hope for clear, informative results overall. Within this broad remit, we can also use factor analysis to get an index scale as part of the overall modelling.

In my example, the proposed asset index becomes one of the S variables, unearned government benefits might also be a simple factor with four benefits (pension, childcare voucher, other medical social security, and housing-cost support), and another factor might be used to simplify the M (memberships) part of the model. One example in recent years is to measure social participation by a series of binaries:

- one binary variable for being a member of a particular kind of social group or civic organisation, including sports groups; and
- another binary for attending that group at least once a week in the recall period.

For 8 civic activities we would have 16 indicators and from this, we may create a single scale.

This would be an example of confirmatory factor analysis.

5.5   Whether to Use Scoping Analysis or Primary Field Research

Another scene where we need the confirmatory approach is in primary data collection with surveys. Whether you write a questionnaire for an evaluation of a service, a study of users' beliefs and values, or to explore lifestyle behaviours and health, or any other carefully honed topic, your questionnaire could contain questions ready for factor analysis.

It will be very wise to pre-vision your need for deep knowledge of scenes from your research topic, and thus stagger your research if you are doing either primary or secondary factor analysis:

- first identify the problem you seek to solve, or work on, and what is already known.
- next identify and carry out the pilot study, interviews, or other active engagement in that scene so that you are more well-informed than ever before to set up the questions or set up the factor analysis of existing secondary survey data. Here there are two different options:
  - Either you are going to do scoping activities such as visiting experts, reading internal documents and reading up via publicly available materials, such as online registers of company accounts or archives; OR
  - You may decide to get ethical clearance to do a more sensitive study which will take longer, involve individuals, and perhaps cover sensitive topics and invoke confidentiality.

It is important to be very explicit about which of these two strategies you have chosen. Within that remit you can then do a variety of things.

The scoping interview is an intriguing alternative to making a larger project. Scoping would refer to visiting an expert in their office or in a public place, or phoning them, and explaining your problem and your research objectives. Ask them to give a view about the situation. See if they can advise you on how wide your sampling needs to be, what pressing issues need to be solved, and how others have been tackling these. Sometimes you find government officers or another organisation are already working on the same problem. It is extremely important to discover what others are doing in your area, so the expert interview is one way to spread your wings into social groups that care about your topic. Lastly, thank each expert and make it clear whether or not you intend to quote them when you write your findings up. They are capable—by virtue of being experts—of saying what they want to say in a quotable way, or explaining to you the niceties of what you should not quote them saying. As you build up your network of experts, you develop a reputation in your field. Be gentle and careful about this. Do not presume that they have free time to give you. Yet it is often easy for professionals who work in office or on-site environments to get 30 minutes free to give you advice on key questions. This is what I mean by scoping with experts.

Mixed methods thus proceeds carefully along the trajectory from setting out the arena of research towards the analytical part of the research. For using factor analysis, a stage of close computer work will be needed. Keeping clear records in a research diary is useful.

Thinking of my earlier 'arena → analysis' diagram (Chap. 3, Fig. 3.2), Fig. 5.5 pinpoints the stages and involvement of expert interviews, primary research, piloting, and other qualitative methods in the factor analysis.

5.6   Research Scope and Feedback Loops

Figure 5.5 illustrates the research design possibilities. It has four panels, for four types of designs but many variants are also possible. These designs are for smaller projects, while the movement from these to writing a whole book might take you into combining projects, along with linking analysis and more synthesis between the projects.

Retroduction is used at two points in this overall process (or perhaps more). First it is used to work out causal mechanisms about which we do not have enough evidence, and thus it guides the search for more evidence. If the latent factor is of a very poor or thin quality in terms of measurement, you may decide to resolve the problem in a two-step process: first choose the strongest indicator among the 2–3 or 4 indicators you have, and drop the others. Your knowledge based on all these is not gone, but your model is going to be stronger for having a single indicator here instead of a weakly performing index. Next, use a generalised linear model or simply clear thinking to recognise that this single 'variable' is representing the whole condition you wanted to represent. In generalised linear modelling, this factor might appear either as an X (independent) or as a Y variable (dependent variable). If it is an X, we are back to the standard regression model with S I M E. When it is Y however the situation gets very interesting.

A generalised linear model creates a two-stage process for representing the latent variable and its causal mechanisms. The latent variable can take one of a wide range of forms, some of which may be familiar to you. The one that corresponds most closely to your simplified factor is probably the normal linear model but the other options are useful, too.

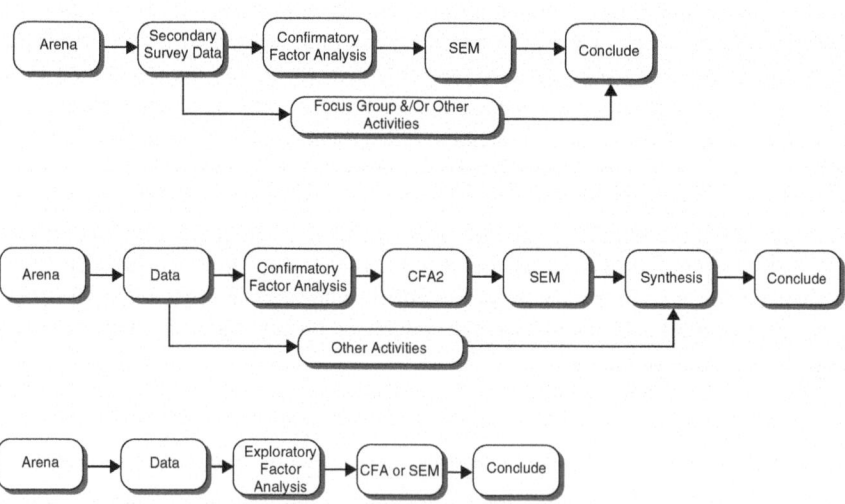

Fig. 5.5 Alternative research designs for projects with index scales. Note: For a small project, use only part of these methods options. Each project will take up just a small subset of these possibilities. Key: CFA = Confirmatory Factor Analysis; SEM = Structural Equation Modelling

– Normally distributed dependent variable. Linear model linked to it by an identity function.
– Poisson distributed count-dependent variable. Linear model linked to it by a log linear function. (A simple discussion of this model is found in Lindsey 1997: 29–30 and throughout.)
– A probability outcome on a 0–1 scale, with a beta distribution representation. Linear model linked to it by an identity function. OR:
– A logit outcome using the log of the odds of Y. Linear model linked to it via an identity function.

These options are discussed in three excellent books, reflecting a blossoming strand of contemporary research in applied social statistics. Dobson and Barnett (2008) discuss a wider range of models, and give examples of computer syntax for estimating them. The latent variable model is made explicit for the probit, logit, and Poisson count models. The correspondence theory of how variables represent real causes in the society is evident in this excellent textbook. Its mathematics is at a level beyond high-school algebra.

Another excellent textbook covering factor analysis integrates structural equation modelling fully (Bartholomew et al. 2011). Here the two-stage mathematical process is presented first, and measures of goodness of fit are discussed in detail in this context. This book is written in an advanced style. Yet these models can be fit using simple software such as SPSS or Stata, too. Whatever your choice of model, it may now become clear that there is an overlap between regression itself (the S I M E approach of Chaps. 1 and 2) and the use of a latent variable(s), since each variable in a regression model is a latent representation of something else in the underlying society. There may also be further overlap between latent variables and the underlying reality. For examples see Byrne (1994, 2011) and Parboteea (2005).

In my own example the use of Likert scales involved switching the signs of some scales and adding a series of qualitative threshold indicators into the CFA. Figure 5.6 shows a situation where there is strong pattern matching among three manifest Likert scales (based on Olsen and Zhang 2015).

5.7 Closed and Open Retroduction in a Factor Analysis Context

Closed retroduction here is likely to involve adding more variables to your model. A project with a factor model may conduct the factor analysis at the beginning or near the end. Either way, the factor analysis can use using randomly selected cases' secondary data or primary data. The scope of the latent factor needs to be expressed clearly in the reports, including every table or figure. For example:

• Note: the population represented are adults age 15–64 in the UK in 2013.
• Note: the population represented are children in India in 2011/12, age 5–17.

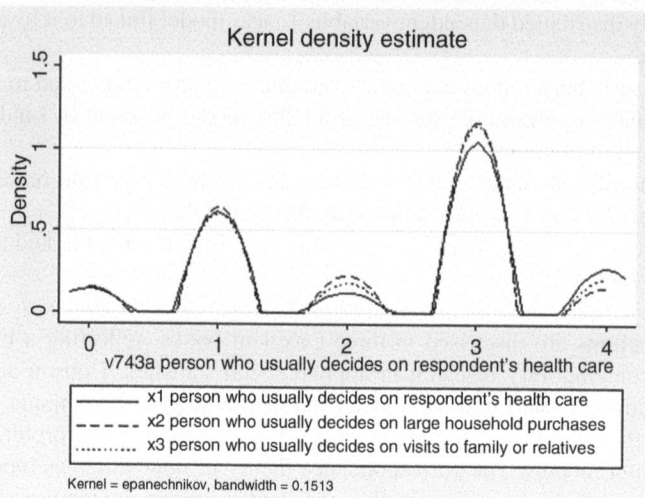

Fig. 5.6 Bangladesh and North Central India, Likert scales of who makes major decisions (raw data). (Source: Demographic and Health Surveys, India and Bangladesh. See Olsen and Zhang (2015)). Here the two countries' data have been merged together, leading to the omission of one 'question' which was omitted in one of the countries' questionnaires. See also 'notes' and 'key' for Fig. 5.3 which used India data only)

These footnotes can be repeated throughout a report to ensure that each table or figure is self-explanatory.

Moving to retroduction, we may ask within the existing data set whether there are additional indicators which might play a role as manifest variables. Alternatively one can develop a regression and thus start a structural equation model. Two roles which the latent factor can play in a structural model are 'independent variable' and 'dependent variable'. I will discuss each option briefly, since this is a potential cause of confusion.

First, most factor analysis studies do not have an explanatory starting point, but rather they set out to describe the latent factors, and work out the number of dimensions N of the factor structure. To begin placing this factor as an explanatory variable in a regression model, you will want to return to the literature review stage, making sure you have covered other key explanatory factors well. Closed retroduction might start to look more like open retroduction because you may need to derive new indicators, perhaps link information in from other sources, and generally open up the research question. Here are two research questions, differentiated clearly on whether or not there is a regression:

RQ1 What factors are associated with higher or lower levels of egalitarian attitudes and gender norms in India among adults living in rural areas, 2015/16?

RQ2 How does the egalitarian gender norm of an area, and couples' deviations from local norms, affect the likelihood of a woman with secondary education taking on paid work of half-time hours or more, in the rural areas of India in 2015/16?

In the first case one might widen the factor analysis to include beliefs about the justifiability of wife-beating, while in the second case, one would be widening it to include all aspects of institutionalised stereotyping, structural relations, and other memberships that might affect women doing paid work.

I have deliberately tried to answer the 'who-what-where-when' questions in framing these research questions. They are also carefully framed to indicate the narrowness of the social scope (rural/urban; region; date) so that there is clarity about the tight framing whilst there is a certain breadth and innovation about the way the question is posed and the variables used.

Instead of the independent variable, your factors may play a role as a 'dependent variable' in a regression or even a QCA analysis. You would look for features of the cases in the data which allow possible deep and proximate explanations of variations in the factor score, within the same geographic space and using the same sample. Keeping to our pluralist stance we might try to gather together one variable representing economic inequality, one reflecting cultural identities, one representing membership in a social group or a scheme, and lastly one reflecting the respondents' age or life stage. The model cannot explain the social norm itself, because this norm is a unitary general feature of the whole scene. A norm is a holistic finding. Instead this regression model is going to try to explain the deviations around this norm: those who are 'less egalitarian' and those who have responded in ways which are 'more or most egalitarian'.

Researchers often check whether the shape of the dependent variable is unimodal, skewed, or bimodal. A histogram can be used for this check. Another step could use QCA (Chap. 6) to check whether some conditions are simply necessary conditions—always co-present—for high or low egalitarianism. Additionally, using statistical methods you can parse the variance of the latent factor into shares, with some of the variance due to each variable. Using interaction effects, you may also consider how the conditioning factors work jointly with the memberships or other causal mechanisms to generate higher or lower levels of Lambda.

Regression after factor analysis takes one of two statistical forms. The first option is a two-step estimation. The factor analysis has its own tests of goodness of fit, and factor scores are 'predicted' for each case. The regression is then run independently. For example with linear regression the project would now be able to have an R-squared value for ordinary least squares estimation. This is suitable for student projects and in the special case of exploratory supplementary localised qualitative or action research. If you can do additional field research, your scope may be quite localised, yet the factor analysis could be national with local-level predictions offering an understanding of how this locality (perhaps one-third of a state in India, around 20 million people; or one region in the UK, containing 7 million people; or a city) is different on average from the national scene. The scope for the further analysis using regression would be more local, so that the regression itself maps on well to action research, stakeholder voices, documentary or interview research within this one locality. These are good rationales for using the two-step process in the statistical stages.

On the other hand, the unitary estimation method would posit the whole model, and enable a single fit of both the factor(s) and the regression. Tests of how well the data fit the model would encompass both the factor model's fit of manifest variables' correlation matrix along with the structural model's fit. Kaplan (2009) explains these tests of model fit in detail.

5.8   The Ontological Element

Theory realism is the phrase Borsboom, Mellenbergh, and van Heerden (2003) use to explain that more than just realism about the entities shown in Fig. 5.1 is required for a coherent approach in cases 4 and 5. They argue that theories must be considered to refer to a real world, and have correspondences to that world, in order for the factor analysis models to make sense. They offer a refined discussion of constructivist approaches compared with realist and instrumental, by which they mean pragmatic, approaches to factor analysis. Realism can help us see that the correspondence of the factor to the manifest variables and to the underlying reality is bound to be imperfect. A realist tends to accept that the world is more complex than the simplified diagram or model that has been posited in this chapter. Therefore, elements could enter in which are neither measurement error nor social deviance, yet are relevant. One could start modelling each omitted factor, and the model complexity zooms up; or one can appreciate that factor models are not going to have a perfect fit in any case. These two solutions have clear schools of thought associated with them in the factor literature. On the one hand, structural equation modelling is a solution to the need to model key processes that might affect manifest variables and cause apparent poor fit. Separate treatments of such models are offered by Bowen and Guo (2012) and McKinnon (2008), implicitly taking a realist approach to the world. On the other hand, if we do not require that we reach a perfect fit, then we may be willing to accept 'error' of fit and realise that it could in reality be parsed into error due to measurement problems, versus error due to other real processes interacting with the indication process. The key conclusion that emerges from both lines of thought is that we do not expect a perfect fit for any particular factor.

5.9   Conclusion

Latent variable analysis enables a research team to summarise a whole lot of facets of a situation with just a few variables. Comparing this result with Chap. 4, we might notice that the Multiple Correspondence Model also involved data reduction, but it did not have the close attention to units of measurement, nor to measures of goodness-of-fit, that were found in the works cited here. Latent variable analysis is a highly developed approach. It can be combined with regression with an explanatory model. You can either try to explain the levels of the latent variables, thus picking out diversity in your sample of respondents; or you can use a latent variable to help with explaining some other outcome. A priori theorising is essential for a successful factor analysis and for structural equation models. Within this scene, mixed methods could play a role at many points. The research design may have feedback loops, but it must in the end synthesise what was found. Otherwise there is a risk of getting lost in detail and not reaching any clear conclusion.

The levels of these variables in sub-groups of the cases can reveal profound social or economic differences. The use of factor analysis in psychology has tended to be more exploratory and in sociology has moved from exploratory methods towards confirmatory methods over the past five decades. Other disciplines that use factor and correspondence analysis include business, geography, health studies, development studies, and education. There is a surge in exploratory methods in data science. For example consumers could be described by a simple pair of two main variables based on a scrape of their web interactions: adventurousness and spending. This kind of simplification may look smooth once it has been done, but the statistical aspects are not simple, and how to add mixed-methods aspects may not be obvious.

A wonderful array of possibilities based on latent variable analysis are available. This array can be daunting to a beginner. Textbooks in this area offer a rich resource, given the ease with which the computer can be used to summarise a large block of data.

Review Questions
Explain the difference between a manifest variable and a latent variable.

In factor analysis, using the structural equation modelling (SEM) approach, the correlation of the manifest variables improves the fit of the overall model. Discuss what is structural about the structural model, giving two aspects of 'structure'—perhaps emergent properties and the fact that a structure cannot be explained purely in terms of its parts.

In factor analysis, the measurement model can be assessed for goodness-of-fit. In your opinion, should a variable be removed if it does not contribute to the fit? Bring in a discussion of prior theory to your answer.

Process tracing stresses process, so give an example of a process of social change and list four kinds of specific evidence you could gather about this example.

Structural equation modelling (SEM) stresses group contrasts. Based on this hint, can you list five ways in which social-class features, both within and across social classes, could be examined using statistical measures and/or survey data.

Further Reading
I recommended using confirmatory factor analysis (Brown 2015). Here, the decision about which manifest variables to include/exclude can rest heavily on theories about reality. These theories in turn can arise from qualitative fieldwork, literature that pre-exists this study, and your theory of change (Funnell and Rogers 2011).

Theory is strongly influential in both confirmatory factor analysis and structural equation modelling (Muthén 1983, 1984). In a complex model, just as in ordinary regression, retroduction can be used to work out—using a mixture of statistical, qualitative, and historical evidence—whether to leave all variables 'in' or to remove one or some of them (Olsen and Morgan 2005). Full models and their variants can be tested as nested models (Bowen and Guo 2012; Kaplan 2009).

In studying norm or the existence of norms, the important thing is that there may be in reality such social norms even though they are not fixed over time, nor consistent across a population. Having started off recognising their potential existence, we can set up measures which tap into norms. Other ways of saying this are that 'we have traces of the real in the data' (Olsen and Morgan 2005); or that we can remove measurement error by using four measures of one thing, as long as they all indicate that one thing indirectly. Both assert 'entity realism'; 'Latent variables are measured indirectly through multiple observed variables' (Bowen and Guo 2012: 18). Entity realism is also discussed by Borsboom et al. (2003). Some textbooks from the 1970s–1990s are out of date or take a different position. With realism, the thing measured is real, and it is not just a variable. The variance around the mode may represent a mixture of real variation, erroneous conceptualisation, and non-cleaned shared measurement error.

References

Bartholomew, David, Martin Knott and Irini Moustaki (2011) *Latent Variable Models and Factor Analysis: A Unified Approach*, 3rd ed., London: Wiley.

Borsboom, Denny, Gideon J. Mellenbergh, and Jaap van Heerden (2003) "The Theoretical Status of Latent Variables", *Psychological Review,* 110:2, 203–219, https://doi.org/10.1037/0033-295X.

Bowen, N.K., and S. Guo (2012) *Structural Equation Modelling*, Oxford: Oxford University Press.

Brown, Timothy A. (2015) 2nd ed. *Confirmatory Factor Analysis for Applied Research*, NY and London: Guilford Press.

Byrne, Barbara (1994) "Burnout: Testing for the Validity, Replication, and Invariance of Causal Structure Across Elementary, Intermediate, and Secondary Teachers", *American Educational Research Journal*, 31, 645–673.

Byrne, Barbara (2011) *Structural Equation Modelling with MPLUS,* London: Routledge.

Dobson, Annette J., and Adrian G. Barnett(2008) *An Introduction to Generalized Linear Models*, 3rd Ed., Chapman & Hall/CRC Texts in Statistical Science, London: Chapman and Hall/CRC.

Field, A. (2013) 4th ed. *Discovering Statistics Using IBM SPSS Statistics*, London: Sage.

Funnell, Sue, and Patricia J. Rogers (2011) *Purposeful Program Theory: Effective use of theories of change and logic models*. Sydney: Jossey-Bass.

Hamilton, Lawrence C. (2006) *Statistics with Stata*, London: Brooks.

Kaplan, David (2009). *Structural Equation Modeling: Foundations and extensions*. 2nd ed. Los Angeles, London: Sage.

Lawson, T. (1997) *Economics and Reality*, London: Routledge.

Lawson, T. (2003) *Reorienting Economics,* London and New York: Routledge.

Lindsey, James K. (1997), *Applying Generalized Linear Models*, New York: Springer.

McKinnon, D. P. (2008) *Introduction to Statistical Mediation Analysis,* NY: Lawrence Erlbaum and Associates.

Muthén, B. & Kaplan, D. (1985) "A comparison of some methodologies for the factor analysis of non-normal Likert variables". *British Journal of Mathematical and Statistical Psychology*, 38, 171–189.

Muthén, B. (1983) "Latent Variable Structural Equation Modeling with Categorical Data", *Journal of Econometrics*, 22, pages 43–65.

Muthén, B. (1984) "A general structural equation model with dichotomous, ordered categorical, and continuous latent variable indicators", *Psychometrica,* 49:1, 115–132.

Muthén, L.K., and B. Muthén (2006) *MPLUS: Statistical Analysis with Latent Variables: User's Guide,* 4th ed. Los Angeles, CA: Muthén and Muthén.

Olsen, Wendy, and Jamie Morgan (2005) "A Critical Epistemology Of Analytical Statistics: Addressing the sceptical realist", *Journal for the Theory of Social Behaviour*, 35:3, 255–284.

Olsen, Wendy, and Min Zhang (2015) *How To Statistically Test Attitudes Over Space: Answers From the Gender Norms Project*, Briefing Paper 3, Gender Norms Project, Creative Commons license, University of Manchester.

Parboteea et al. (2005) "Does National Culture Affect Willingness to Justify Ethically Suspect Behaviours?", *International Journal of CrossCultural Management*, 5:2, 123–138.

Potter, Garry (2000) *The Philosophy of Social Science New Perspectives,* Essex: Pearson Education Limited.

Ragin, C., (2009) "Reflections on Casing and Case-Oriented Research", pp 522–534 in Byrne, D., and C. Ragin, eds. (2009), *Handbook of Case-Centred Research Methods*, London: Sage.

Sayer, Andrew (2000) *Realism and Social Science.* London: Sage.

Qualitative Comparative Analysis (QCA): A Classic Mixed Method Using Theory

Qualitative comparative analysis (QCA) and fuzzy set causality analysis are comparative case-oriented research approaches. The basic technique was developed by Charles Ragin, with key presentations by Rihoux and Ragin (2009), Thiem and Duşa (2013a, b), Schneider and Wagemann (2006), and Rihoux and DeMeur (2009). A case-study comparative analysis will have a series of cases, all somewhat similar or comparable, and may use a variety of methods to make comparable measurements or record qualitative data on all or most of the cases. In a sense this kind of study might be thought of as the creation of a reasonably small matrix of both qualitative and quantitative data. Some examples can be seen in Appendix Table 6.2.

Methods that use the mathematics of fuzzy sets have shown deep linkages between the real world represented by the data and the mathematical manipulations that can show us patterns in the QCA- and fuzzy-set-type data. On one hand some of the patterns may reflect underlying causal capacities, mechanisms, or pathways in which an 'X' explains the occurrence of a 'Y' (Ragin 2008). On the other, the patterns might be seen as a broad representation of social structure without implying one-way causality (Smithson and Verkuilen 2006). In Fig. 6.1 I illustrate how the approach refers to the real world via evidence that is put into the empirical sphere. The three spheres in this diagram are the real, the empirical, and the actual. The point of having a three-sphere approach is to realise the great weaknesses of the empirical data: they may be incomplete, badly presented, or even misleading. The actual world, however, is so vast that we routinely have to reduce it to a set of evidence, and thus to matters of record that are transparent and ready for a concrete and reasoned discussion (Goertz 2017). I believe there are wide-ranging implications of the debate about QCA: it is not just a narrow debate for technical experts. For instance the methods of 'a frequentist regression' and QCA might reach vastly different conclusions about a medical treatment or a social intervention. The applications of regression and QCA overlap, and they include all the disciplines ranging from management and public policy to medical, nursing, and demography, and many others.

W. Olsen, *Systematic Mixed-Methods Research for Social Scientists*, https://doi.org/10.1007/978-3-030-93148-3_6

This chapter shows how the mixed-methods debate feeds into the measurement issues, the pattern discernment stage, finding out about how real causes work, and drawing overall conclusions from QCA studies. The debates about validity are not just about measurement, but about the reality behind the data. Therefore the arguments in this chapter are immune to several key 'rebuttals' of QCA which were grounded in methodological individualism (Lucas and Szatrowski 2014) or deductivist methodology (Baumgartner and Thiem 2017). I will discuss the differences of standpoint that have emerged during the debates about case-study analysis.

Useful reviews of the whole QCA method are found in works that cut across disciplines such as Ragin (2006), Vatrapu, Mukkamala, Hussain, and Flesch (2015), and Byrne (2005). I summarise the realist approach using simple diagrams (Olsen 2014).

Comparative case studies are an extension of the single case-study method to examine the systematic analysis of patterns in groups of cases. Case-study research can involve qualitative in-depth investigation of how similar events are caused (often also involving some multilevel process tracing; see Bennett and Checkel 2014; George and Bennett 2005). Ontological depth in the casing is commonly one aim of the exploratory stage of research. For example Lam and Ostrom (2010) looked at watershed sites in Nepal, along with doing a field study of households who used water for both consumption and agriculture. The systematic analysis approach known as qualitative comparative analysis (QCA) and fuzzy-set analysis (abbreviated fsQCA) use a range of binary indicators and ordinal rankings to draw contrasts (Rihoux and Ragin 2009).

▶   **Tip** QCA usually uses qualitative information about the cases. Clusters
    of cases, known as configurations, are shown to have relevant similarities
    with regard to the association of their contexts or characteristics with
    some key outcome. This logic is similar to regression. But in QCA, the
    causal mechanisms are not assumed to work identically across the
    whole world or 'population' of cases.

Mixed methods is commonly the fundamental basis of QCA. Fieldwork and documentary research are commonly used (Ragin 2006). One of the methods that helps with this kind of complex research process is fuzzy-set measurement. Fuzzy sets measure the degree to which a case meets the criteria for membership in a qualitatively defined (or simply real) set. QCA brings together the ideas of the case, its features, other cases which are either nested within, or non-nested relative to the original type of cases, and measurement methods along with ideas about causality.

Qualitative comparative analysis (QCA) opens up two new forms of knowledge: firstly knowing about alternate pathways to one outcome (equifinality) and secondly perceiving nuances of necessary cause and sufficient cause. These gains in knowledge comprise part of the added value in the QCA method. We can have these gains, whether or not we use fuzzy sets. Compared with fuzzy sets, a simpler measurement metric is 'crisp sets', which are binary variables reflecting nil or full set membership (Caramani 2009). In some ways, fuzzy set analysis is consistent with statistical analysis, and both are certainly amenable to using mixed methods across the qualitative-quantitative divide.

In this chapter I will present only the basics of QCA because there are excellent books on it (Ragin 2000, 2008, Rihoux and Ragin 2009, and Ragin and Davey 2014; or Kent 2007, a chapter on QCA). Applications of QCA often include holistic variables. Policy environment, regulations across a wide international comparative scene, or local cultural differences all fit in well with non-individualistic QCA (Rihoux and Grimm 2006).

One can use just part of QCA without harming your research design; it is not necessary to apply all the tools in the QCA fuzzy set toolkit. Therefore, I start first with what the 'truth table' achieves. It is a descriptive table offering a platform for interesting, expansive interpretations, and perhaps inductive generalisations. Then I look at the reduction of the truth table, a systematic method of transforming your data using fuzzy-set logical rules. This is based heavily on Boolean algebra so I set out what these rules are. Applying them can lead to anomalies compared with standard statistical reasoning, but it can be stimulating to compare the two styles of reasoning.

The last part of this chapter sets out consistency tests of how well a QCA model fits the data. It is possible to imagine a future expansion of the methods in this area. I set out some methodological questions which can now be answered in fresh ways. It all boils down to realism again, in this sense: for three levels of reality (empirical, actual, and 'the real') there are three uses of QCA—truth tables, examining causality via Boolean reduction of the causal pathways, and discerning the real causal mechanisms with mixed methods (see Chap. 1 for three levels of reality; and Olsen 2019).

Combining knowledge at all three levels will be useful, but it is not necessary to use 'reduction' nor 'parsimony' (according to my arguments), so my advice diverges from key competing texts (Lucas and Szatrowski 2014). The main reason is that there is real causal complexity in the world. In the introduction I will spell out an early stage of a study of child labour to illustrate how realist triangulation would be different from the reductionist approaches (which include both QCA reductionist methods and statistical reductionist methods).

In both cases, the supposed tendency to reduce the model to a parsimonious model only gives you gains in certain directions, and it turns out to be a preference based on the wrong reasoning. By contrast, I hold values that pull me in the other direction towards pluralism and multidisciplinarity. In the following chapter, I spell out some ways in which the realist approach can serve very well the researchers' needs. In Chaps. 8 and 9 I explore more how epistemological values influence the triangulated, mixed-methods project.

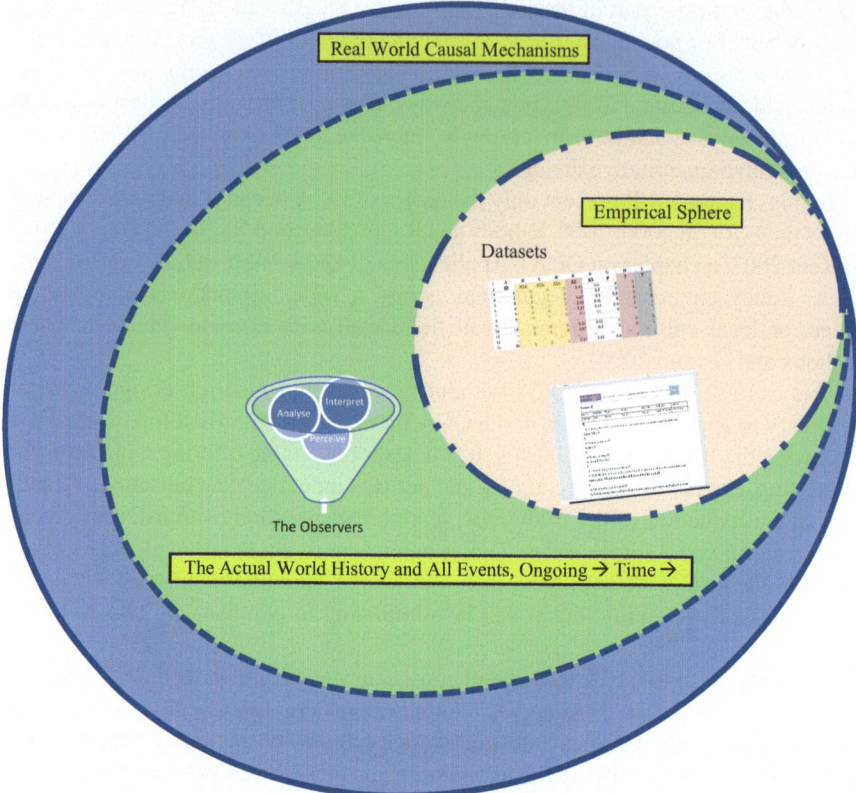

Fig. 6.1 Three spheres in realist qualitative comparative analysis (QCA)

6.1   QCA Is an Umbrella Over Many Procedures

Many different kinds of studies qualify as a comparative analysis using qualitative data. For them to fit in the QCA umbrella it is usually dealt with as shown in Figs. 6.1 and 6.2. In Fig. 6.1, crisp sets have been used to summarise a dataset, and the patterns shown have an independent variate on the horizontal axis (i.e. seen in the columns), and a dependent variable—the outcome we are focusing on—in the vertical axis (here the rows).

For example if X was sufficient to cause Y then every time X is present, Y is high (Evidence Pattern 1). The figure also presents a less clear pattern at the right-hand side. In Evidence Pattern 1, each time a case appears in the bottom right cell, it contradicts the hypothesis that X is sufficient for Y. Using a mixture of crisp and fuzzy-set memberships, we have empirically discerned sufficient configurations using data on slum-dwellers in Chennai (Harriss-White et al. 2013). We found a lot of variety among the slum-dwellers and among their households, so that the effects of a great recession and crisis (X) were not having clear, undiluted effects on their overall well-being (Y). In fact it was clear that a wide range of demographic and

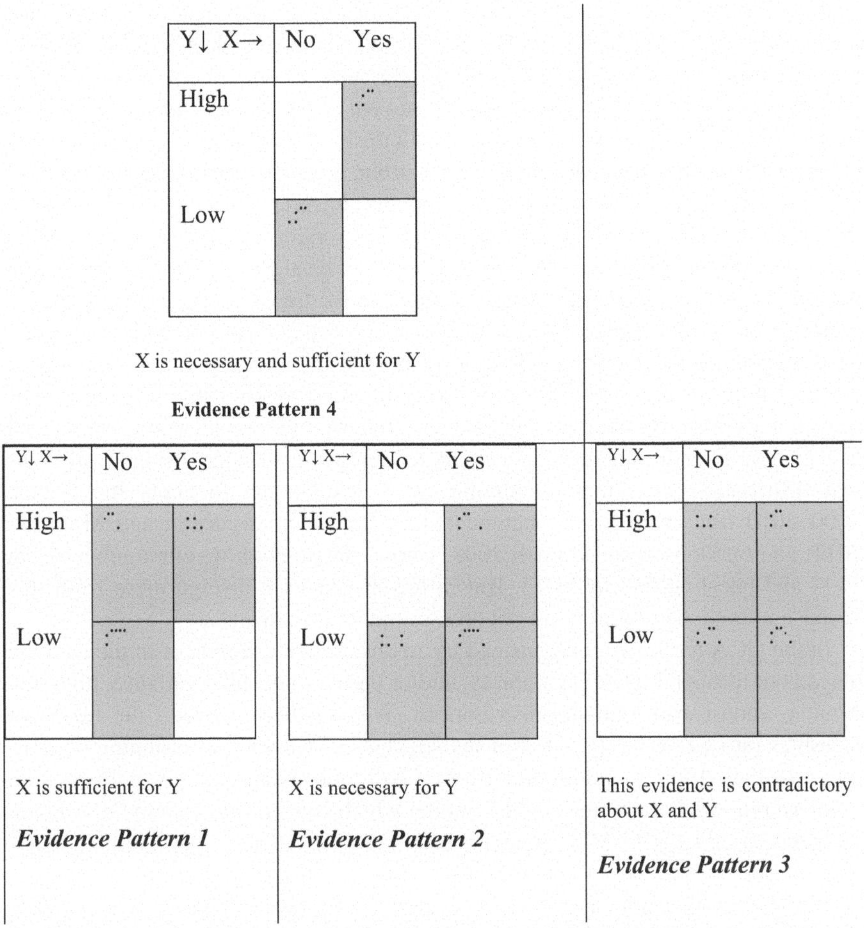

Y↓ X→	No	Yes
High		.··
Low	.··	

X is necessary and sufficient for Y

Evidence Pattern 4

Y↓ X→	No	Yes
High	.··	::.
Low	:···	

X is sufficient for Y

Evidence Pattern 1

Y↓ X→	No	Yes
High		.·.
Low	:. :	:···.

X is necessary for Y

Evidence Pattern 2

Y↓ X→	No	Yes
High	:.:	.·.
Low	::·.	:·.

This evidence is contradictory about X and Y

Evidence Pattern 3

Fig. 6.2 Four patterns and their interpretation in qualitative comparative analysis. (Source: Olsen 2014)

economic factors could intervene to ensure well-being in spite of the crisis. QCA, on the other hand, is meant to allow for these simultaneous factors.

QCA is not intended to measure a dose-response type relationship, or to make exact predictions of the size of the impact of X on Y (top row of Fig. 6.2). A dose of medicine X could cause an improvement in fever Y, which in QCA terms would be described as follows: X is not a necessary cause of lowered fever, since time passing and the immune system together may be sufficient and not require the medicine. On the other hand, X is a sufficient cause of lowered fever (written X => ~Y meaning X is sufficient for Not-Y). The evidence to support this hypothesis would be found in Evidence Patterns 1 or 4.

The use of a simple X-Y diagram understates the scientific step-change that Ragin and his colleagues have created. The X and the Y can, in turn, represent complex combinations of social and other conditions. Therefore we need to test a myriad of possible X combinations and several options for Y. To illustrate, we just saw that the causes of Y and Not-Y were not the same. X in turn could be medicine+immune, medicine, immune, and so on, until we have listed all the possible combinations of all the underlying X conditions. Ragin stressed that these conditions would include not only personal traits (such as immunity or dose) but also social and historical contexts (Byrne 2005). This approach is consistent with the harmonious use of quantitative and qualitative reasoning advocated by Bryman (1996, orig. 1988) and Olsen (2012).

To preview Chap. 7, have set out first how the conversion of ordinary variables to fuzzy sets is carried out. There can be direct conversion to crisp or fuzzy sets for binary or ordinal variables, respectively. There can be algebraic indirect conversion for continuous variables; and there is the option for transforming and shaping the scales in fuzzy-set membership space so that they are comparable with each other, or so they match the data better. This latter method is known as calibration. Using a mixture of conversion and calibration, you can create variates that fit your modelling strategy well. By contrast, in basic statistics we often seek to ensure a normal distribution. Even in statistics, the general tendency has changed; using mostly normal distributions was popular among some scholars up to about the decade 2000–2010 (the change was documented in Bartholomew, Knott and Moustaki 2011; and by Dobson and Barnett 2008, whose classic book was first published in 1994 and latest edition in 2018). But in recent years we diverge more from this because normal distributions are not an essential requirement.

In the QCA toolkit we may instead try to ensure that differences in the data are played out through high comparability, and/or perhaps also high variance, high dispersion, and/or even bimodal distributions. We can then reconvert the fuzzy-set measures into a Z score space to run statistical tests—all without assuming normality of the data (Eliason and Stryker 2009). I will look at causality given this system of measurement. Beyond the S I M E model which converts easily into QCA logic, we need to add the C M O model which offers awareness that Context (C) + Mechanism (M) produces Outcomes (O).

Whilst I offer a summary of the CMO model here, I explore more of its implications for how we study obstructions and enabling causes in Chap. 7. We find that 'interaction effects' in regression are a dim and overly slimmed-down way of moving towards the QCA causal pathways approach. The two are complementary, so I recommend, at the end of Chap. 7, the combination as follows: Arena setting, refining the measurements, QCA, then statistical analysis, and interpretation, with iterative feedback loops, before making the final interpretation. For a detailed summary of such a logic see also Danermark et al. (2002). Danermark et al. were writing about research projects in general, and advised that they have an iterative series of steps with feedback. This works well in a realist QCA framework.

6.2   Tables Help to Summarise Qualitative Comparative Evidence

The QCA method—as currently used—is deliberately always qualitative. Over 950 studies have been gathered by the learned society that promotes QCA, 'Compasss' (www.compasss.org). Without the qualitative research foundation it becomes mainly a mathematical system (as seen in Mendel and Korjani 2012, 2018; also seen in Baumgartner and Thiem 2017). QCA founder Charles Ragin has argued the case for intensive qualitative research using case-comparative methods for over 30 years (Ragin 1987), yet we do also see numbers in the tables of data used by this school. Thus it is not exclusively qualitative. There is a deep linkage of the numbers with the knowledge about the cases and the configurations (groups or types) of cases.

In the example shown in Fig. 6.3, which illustrates the key causal analysis by Ragin (2000), a spreadsheet table offers access to the qualitative data via a much-simplified summary. The data are provided in Appendix Table 6.3.

In this table we see five explanatory factors which I will briefly explain, and the outcome is the tendency of a country to have had riots and street protests against the International Monetary Fund (IMF). The time period covered is 1972–1990 and the countries include 57 cases, constrained by the research time available. The qualitative research involved studying online and paper news reports of protests and riots, along with triangulated use of government and World Bank data on economic flows. The first variable found to negatively affect protest was having a high level of government spending and activity. It is a 'crisp set' (high or low represented as 1 or 0). Two more

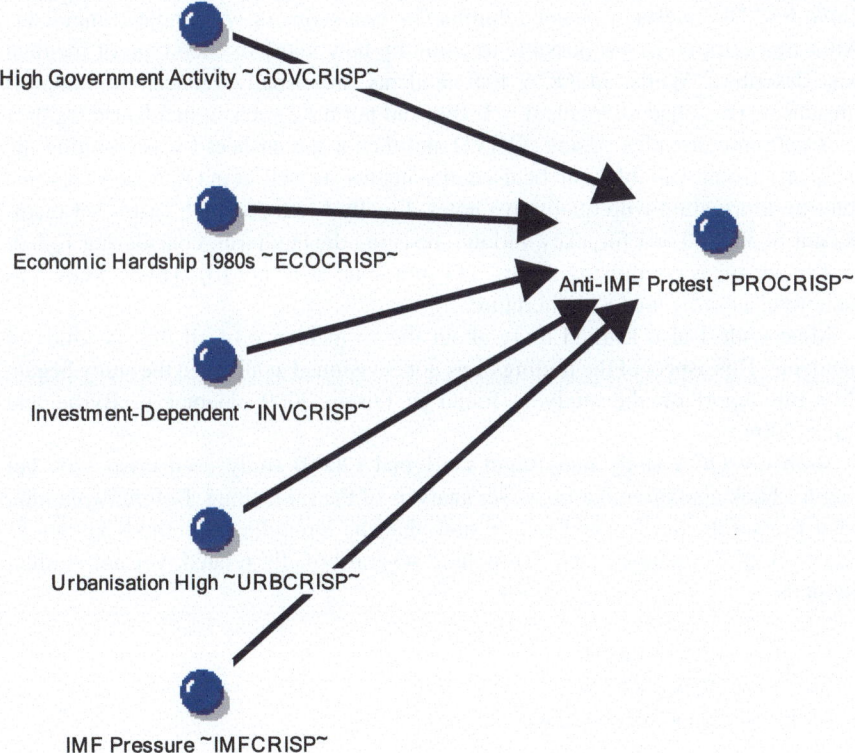

Fig. 6.3 Explanations for anti-International Monetary Fund protests. (Source: This figure is based on analysis of the data and theory in Ragin 2000)

economic variables were tested for positive associations with protests: economic hardship and poverty (denoted by Ecocrisp in Fig. 6.3), and being dependent on foreign direct investment to a high degree. Next a demographic variate was created to reflect the degree of urbanisation, because as Ragin (2000) found, countries with large rural populations had less tendency to protest. Finally, one necessary cause was discerned: the country could not have protests against IMF unless it already had IMF intervention and policy changes resulting from IMF pressure. Typically IMF pressure involved conditionality on an IMF international loan. The IMF would loan funds to a government but in exchange, the IMF and other international donors would jointly require that government to privatise the national organs such as electricity and gas, and deregulate food markets and trade. It was not enough simply to repay the actual IMF loan. The whole process became known as structural adjustment.

The findings from this study were clear: for the countries involved, IMF pressure was a necessary but not sufficient cause. Further political economy mechanisms were reflected in the other four variables, which played a part in several pathways to protest. These pathways were each 'quasi-sufficient', which means sufficient but not deterministically always having the same effect.

To illustrate the creation of crisp sets, I used a series of 39 interviews to generate case-study data in a comparative table. One part of this table is shown in Appendix Table 6.4. The interviews gained comments on a series of work-related incidents. After transcription it was possible to count up how many of each type of incident was described. Whilst in QCA the incidents are usually reduced to crisp-set 'instances' (case had an incident = 1, case did not have such an incident = 0), they start out more complex. Using NVIVO and then a spreadsheet I was counting the incidents. Count variables can be used in statistics, as well as in QCA, so I kept this interim information with qualitative notes. Finally I reduced each case's set membership score to 0 or 1 for a key variate, 'resisting the landlord', considering 1, 2, 3, or 4 incidents as comprising a 'yes' (1), and none as a 'no' (0). This is known as dichotomisation in the QCA literature.

Meanwhile I also learned more about the co-incidence of all the six kinds of incidents. This aspect of the findings was not envisioned at all when the study began. (My full report on the study is found in Olsen, 2009, chapter in Byrne and Ragin 2009.)

Both my QCA study and Ragin's original (2000) study used crisp sets, but Ragin's book also offered a fuzzy-set analysis of the same thing. Before explaining what fuzzy sets are, I would like to note that the umbrella of methods known as fuzzy-set QCA (denoted fsQCA) include several broadly related, yet not similar, methods:

- Crisp-set QCA (csQCA)
- Fuzzy-set QCA (fsQCA)
- Multi-valued QCA (mvQCA)

For a taxonomy of the various QCA methods, see Rihoux's overview in Rihoux (2006). These methods are generally used by mixing a qualitative basis in policy and historical research, sometimes field research and individual case studies, tabulation of both aggregated and holistic variables—such as investment ratio and the IMF pressure in the above example—and finally a data reduction to find out key pathways of cause. In every example of fsQCA the discernment of necessary causes is an important first step. Necessity and sufficiency of cause do not have the same meaning.

A fuzzy set is a membership score between 0 and 1 which reflects and represents how closely a particular case conforms to a defined case characteristic. Thus a fuzzy-set variate is a measurement vector (a long list of numbers), each between 0 and 1. Fuzzy-set values of 0 mean the case is not at all in the set. Those measured at 1 are fully in the set. Sometimes to keep it simple the researchers avoid fuzzy sets and use crisp sets only. A crisp-set summary of a fuzzy set might simply round down the fuzzy-set scores of 0.499–0 down to zero, and round up the rest of the scores from 0.5 to 1 to '1' ('in' the set). Other methods of equating a fuzzy set to a crisp set, or vice versa, are known as calibrations (Chap. 7).

Fuzzy-set QCA is a promising and well-established systematic case-study method. Examples with datasets can be found in the JISC online email list about fuzzy sets and QCA, QUAL-COMPARE (www.jiscmail.ac.uk), or in the bibliography on the COMPASSS web site (sic) www.compasss.org, which is dedicated to the needs of those who do research using small and medium numbers of cases. A crisp set is a 0/1 binary which indicates whether or not a case is in a set, and crisp set QCA is denoted csQCA.

Fuzzy-set social science and csQCA, coming under the umbrella of QCA, involve research designs that explicitly delineate a series of cases as single-level or multi-level, nested or non-nested, units. Postal services, for example, include the government Post Office plus a variety of parcel delivery service companies. Bank customers include organisations and individuals. Case-based researchers make a virtue of comparing cases of varying types and sizes. The 'casing' stage of the research involves discerning the cases (Ragin 2009).

It is possible to show linkages between quantitative levels of measurement and fuzzy-set scorings, as shown in Table 6.1.

Table 6.1 Level of measurement and creation of a fuzzy-set membership score

Variate level of measurement:	Binary	Ordinal	Continuous
Parent's education:	Literate or not	Level of school completed	Years of schooling
Child labour:	Whether the child's work exceeds the hours/week cutoff	Hours worked relative to the threshold: Below, Above, Above Adult Threshold	Hours worked over the age-related hours/week threshold
Assets:	Roof is built of solid material	The degree to which they have any of these: solid roof, piped water, toilet, electricity, vehicle	Factor scale (known as an asset index)
Fuzzy-set comparator:	Crisp set	Possibly use a multi-valued set	Fuzzy set
Alternative:		Could use 3–4 crisp sets	

A	R	F	L	C	W
0	1	0	1	1	1
0	1	0	1	1	0
0	1	1	1	1	1

Etc. for a total of 19 rows, representing 19 watersheds.

Fig. 6.4 A QCA data set for watersheds in Nepal with crisp variates only. Where (A) represents whether a system has received infrastructure assistance since the competition of the WECS/IIMI project. (R) represents whether farmers on a system have been able to develop a set of rules for irrigation operation and maintenance. (F) represents whether farmers have worked out provisions for imposing fines. (L) represents whether leadership in a system has been able to maintain continuity and to adapt to chaining the environment. Where (C) represents whether farmers have been able to maintain a certain level of collective action in system maintenance. (W) is equal to whether watershed has a good water supply. (Source: Lam and Ostrom 2010).

It is useful to glance at a micro-level QCA data set to see how easily it can be coded into a spreadsheet format. See Fig. 6.4 which has only one level of detail.

To become a 'truth table', this table could be collapsed slightly and a column for 'N = Number of cases' added. Then we seek evidence that the pattern of the data fits one of two hypotheses about causality.

The Online Annex shows the summarised truth table and the complete algebraic model statement (https://www.github.com/WendyOlsen/SystematicMixedMethods). This conversion of data to a Boolean statement is rather technical but also brings insight into the asymmetric X-Y relationships.

The Boolean summary is an expression of the data, and thus summarises the evidence. However, Boolean summaries can be reduced by a process of Boolean reduction. The initial summary is complex, with one term per sufficient pathway to Y. Then a simplified set of terms may be produced as an intermediate or parsimonious re-interpretation of the data. QCA offers a logic around this kind of simplification of data tables. The controversy that critics of QCA have raised centres on whether the Boolean reduction is a valid representation. I have left this matter for other venues, focusing in this book entirely on the 'complex' or complete representation.

For mixed methods, one may also decide to move in a different direction. You may decide to use a variety of statistics, descriptives, graphics, quotations, or case-study stories to illuminate examples, which are part of what is shown in the table or figure of QCA results.

Every single case is very important and the sampling or selection of cases will influence the conclusions, as noted by Goertz (2017). For instance, Goertz refers to mediating causal mechanisms, M, as follows: 'Causal mechanisms in multimethod research practice are almost always significantly more complex than the simple X → M → Y' (Goertz 2017: 36). He goes on to discuss the cross-effects of combined causes, the overlap of causes through a causal chain, and the resulting issues for generalisation and robustness.

6.3 Data Reduction Has Been Well Theorised

Many practitioners of QCA have reduced their truth table to an algebraic statement, which summarises the configurations in a series. They use + for 'OR' and * for 'AND' to help these representations.

After representing the material succinctly, it is then possible to reduce the configurations to a simpler set of terms. However it is important to realise that the practitioners quoted in this book all would concur with how Snow and Cress saw QCA:

> QCA, ... is conjunctural in its logic, examining the various ways in which specified factors interact and combine with one another to yield particular outcomes. (Cress and Snow 2000: 1079)

The way we see the analysis is that each conjunction is contingent on the history that led up to its measured features. The claims made are not lawlike claims, nor are they meant to imply universal validity, or any validity outside of the time and place that are being referred to. The claims are definite with respect to the past/present as it was studied; but conjectural and contingent with regard to the future. This idea that causality can be firmly believed in, yes one could nevertheless also accept the complexity-theory approach to causes, requires a sophisticated idea of 'cause and effect'. The ideas we have about this underpin our use of 'reduction' to simplify how we represent the data table.

One simple summary of how we see this is that causality is not deterministic. A particular cause cannot work the same way every time in society because the historical circumstances are different. In other words, other things get in the way. Yet it has that tendency to cause that effect which we are going to describe. Fights cause divorce, drinking causes liver problems, trouble at school causes trouble sleeping at night, and so on. We mean that the cause tends to have its effect.

A second way to summarise this is known as 'context-mechanism-outcome' (CMO) approach. Introduced by Pawson and Tilley (1997) and promoted by Befani et al. (2014) and Cross and Cheyne (2018), the C refers to Context, M to Mechanism, and O to Outcome. The point is made that the outcome depends not only on one mechanism, but also on the several contextual factors which form the context. As we move forward in time to new situations, the context may be quite different so the mechanism will in fact have a different effect. Yet it can be shown, with strong validity, to have its effect across a wide spectrum of cases, as illustrated in many studies. In this volume the CMO approach is analogous to the S I M E approach except that the acronym S I M E leaves out the explanand, and focuses only on the causes. CMO argues that effects (labelled O for Outcome) depend on a mixture of contexts and mechanisms, while S I M E is meant to argue that the mechanisms themselves are of many different types.

There are two methods of data reduction in wide use today. The first, led by Ragin, is to apply laws of logic and deduction to simplify the truth table. Examples of this method include Lam and Ostrom (2010), Vis (2009), and Cross and Cheyne (2018).

The second method is to seek the necessary causes and then the sufficient causes. We also need to look for the causes of the negation (and what happens if there is negation of a cause). In other words, to follow a truth-table protocol. Ironically, the first method has a strong deductive character, and the second method has an inductive character. Ragin and the other authors all use the two methods side-by-side. Their protocol is well known. A clear, detailed version of it is provided in Appendix Table 6.4. The overlap of these two steps—necessary and sufficient causality—helps reinforce my claim that these varied logics are reasoning moments, rather than social-science methodological schools.

It is unfortunate that some critics of QCA, whom I have cited, ignore the necessary cause part of the argument, and neglect that part. They spend much time critiquing the sufficient cause part without realising that the authors are fundamentally not making the same kind of claim as a statistical claim that 'X is both necessary and sufficient for Y'. Nor is the QCA expert describing a permanent, reproducible relationship like a dose-response function. A dose-response function is an X-Y plot showing how more or less of treatment X gives rise to higher/lower levels of outcome Y in regular, reversible ways: more dosage, more of Y; less dosage, less of Y and so on. In QCA there is no assumption of reversibility. We say that QCA does not assume symmetry: it finds whether X caused Y in the data, but it would be a different test to find out whether a reduction in X caused a reduction in Y.

QCA does not assume closure. Regression aimed at estimating a dose-response function assumes closure.

Here, I will spell out the protocol first, then explain the logically deductive simplification method.

First we test each column for necessary cause. I present this as whether each X influences the outcome Y. Secondly, we begin to compare mixtures of conditions, permutations, and combinations, all of which together are known as the 'conjunctions' of the conditions. The conjunctions include not only direct causal variables, but also the influences represented by proxy variables and the helpful impact—or barrier—offered by specific conditions. It might be best to relabel these three kinds of conditions as X, P, and C, referring to causes, proxies, and contextual conditions respectively. Normally we do not categorise the variates so simply though. It is worth pointing out, too, similarities between the S and I part of S I M E, with structural and institutional variates being part of the context; yet these may turn out to be causal or proxies once they have been operationalised.

To 'operationalise' commonly means at the very least to measure something. Operationalisation in general refers to how we systematise a whole table based on a range of relevant (plural) theories of what causes the outcome. Therefore operationalisation is not just one variable at a time, but needs to take into account minimising overlap between the variables. If we have a direct measure, we do not use a proxy, because no new information is added by it. If structures of social class have direct effects, we consider a class variable as a causal mechanism; if they are also contextual then we allow the same variable to have a contextual effect. Therefore, in spite of my fine distinctions between causes X, proxies P, and context variates C, we don't use these labels explicitly much in practice.

There is a big difference between 'testing for necessity' and 'testing for sufficiency' as Ragin has laid these out. Testing for necessity refers to necessity across the whole range of cases represented by our data set. By contrast, when testing for sufficiency we might be satisfied if a causal conjunction was working on just 10% of the cases, or some subset whose size we do not know in advance. This special treatment of subgroups in QCA is a special advantage of the method. I will be talking about 'permutations' as a way to get at it. To summarise:

- Necessary causes in general are the ones that apply to all permutations.
- Sufficient causes are those which work along pathways, either common to all or specific to a subgroup of the permutations.
- Cases lie in permutations, because their set membership scores show which conjunctions they are in.
- Evidence supporting 'necessity' of X for Y has an opposite triangular appearance compared with evidence supporting the sufficiency hypothesis.

A permutation is thus defined not only as a set of cases, but as a set of conditions which apply to a group of cases. A configuration is a set of conditions combined with the actual observed outcomes for that permutation.

Thus on one side we have X, P, C and so on, and finally at the right we place the Y variable. If one of the causes is necessary across the whole scene, then we do not need to include it in the tests of 'sufficient causes'. The reason is simple: it is found in every causal condition set that had a positive outcome!

Ragin calls some particular cases instances. A case is an instance if it has a positive outcome. A permutation that has all positive outcomes is a configuration with consistency 1 (one) for the claim that X causes Y. Thus I have now defined consistency as a number valued from 0 to 1 which measures the extent to which the configuration has the outcome pattern and the underlying causal pattern that is associated with 'sufficient cause'. At this point it is crucial to realise that in Figs. 6.1–6.2, I have used X to refer to the whole fuzzy-set intersection of the variates in one permutation.

If the data holds M configurations, then we may have M permutations and M possible x-y diagrams to illustrate the situation. I have depicted all these M possible x-y fuzzy-set graphs in the software 'fsgof' which is based on spreadsheets. The actual number observed may be less than the theoretically possible number M, and this situation is called limited diversity.

I would like to summarise, then show two examples. First, Ragin showed, if all the Ys have a particular X attribute, we might call this one A, and then it can be argued that A is necessary for Y to have occurred. We can remove this necessary variate from further analysis, but return to it in the interpretation. Each X is checked for whether it is necessary; so the necessary causes may end up being A B and C.

Secondly, Ragin urges us then to look for causal pathways. These are sufficient causal combinations. This is done either one by one, looking at the consistency of configurations, or using fsQCA software by Boolean reduction. The full explanation of Boolean reduction is beyond this book, but is covered well elsewhere (e.g. Schneider and Wagemann 2012. Analysts using R software will appreciate Thiem and Duşa (2013a, b). Those who use fsQCA will want the guide written by Rihoux and Ragin (2009), whose Chap. 3 covers crisp sets and Chap. 5 covers fuzzy sets.

Thinking of the crisp-set situation, each permutation may have either one case, or more than one cases. Those with more than 1 case can appear contradictory: some cases have Yes on Y and other(s) have No on Y. The tricky thing about a crisp-set contradiction is that we might decide to ignore it and move to other permutations. Software allows you to omit some permutations by stating 'omit' or 'disregard' these permutations. That might be the case if you are aware of something particular about this group of cases. However if you don't want to ignore it, the case(s) that have No on Y will be very powerful in rejecting the sufficiency of X for Y.

We are looking at Evidence Patterns 1 and 3 in Fig. 6.2. Therefore, two preconditions must be met for the analysis to go further, with regard to this permutation. First, there must be variation on X: some cases have it, some cases do not have it. Without variation on X, we cannot tell what effect X has, and no conclusions will be possible about X's role in causing Y. Second, there must be variation in the Y outcome; without any instances of Y in this group of cases, we cannot test for 'sufficiency', because we cannot tell what caused (or was associated with) Y.

Because of these basic, straightforward rules of interpretation, one learns a lot about sampling theory by examining data using a QCA lens. One begins to realise that the sample coverage must be large: high and low values of each X and of Y. The sample implementation should reliably generate a selection of cases from the full extent or range of the distribution of all possible X and Y values (based on realistic, evidence-based expectations). Without such a good sample, the conclusions are going to be weak.

We can test a variety of X hypotheses, with X being a single variable or a multivariate vector, such as here ARF is sufficient for good water supply W. If any or all of these is necessary for Y=W=WATERSHED HAS GOOD WATER SUPPLY then we say X is necessary for Y.

The computer or the researcher can look for sufficient pathways that meet this requirement: If in all cases within a configuration, those which have high levels of X have just as high levels of Y, the pattern suggests that X is sufficient for Y (see Olsen 2014. Counting up these cases, we also standardise by adjusting for the size of the whole sample and the size of the group that does have Y greater than or equal to X—in the set showing inclusion of X in Y. Simple Venn diagrams can help reinforce the point that sufficiency is when there are no cases where X occurs without Y. (By contrast, for 'necessary causality' of X for Y, we have the converse, no cases where Y occurs without X.)

As a second example, I will set out the difference between necessary and sufficient causes using the final conclusions of Vis (2009). The research question was: 'Under which conditions do governments engage in unpopular social policy reform and when do they refrain from doing so?' (Vis 2009: 42). Among the three hypotheses she looked at closely was this one: 'H1: Governments pursue unpopular social policy reform only if the socio-economic situation and/or the cabinet's political position is deteriorating' (ibid.: 37). Vis's exemplar is useful because it also illustrates the testing of a hypothesis regarding the absence of a specific, carefully defined outcome:

> H2: Governments shy away from pursuing unpopular measures if the socio-economic situation and/or the cabinet's political position is improving (ibid.).

The QCA analysis of hypothesis 2 would not be a simple converse of the findings for hypothesis 1 because of asymmetries in the Boolean algebra and the careful treatment of necessity separate from sufficiency. Vis, like others, focuses mainly on what Xs are sufficient for one key Y outcome. Her main finding can be summarised as follows. It is a sufficient pathway for unpopular reform to occur [in this study's scope] to have both a weak socio-economic situation (denoted WSE below), and also one of the following two factors: either a weak political position (denoted WPP) or a rightist government (denoted RIGHT). In her results section, she draws attention to a 'necessary' cause too: 'a weak socio-economic situation is necessary, but not sufficient, for unpopular reform' (ibid.: 45). This paper is a good illustration then of the difference between the two.

It is possible to use Boolean algebra to represent the findings insofar as they pass the consistency cutoff level. Vis (2009) summarises her findings as shown here:

> WSE*(WPP +RIGHT) → UR, where WSE refers to one configuration, WPP is another, RIGHT is a third, and UR is the key outcome.
> (consistency: 0.90). (Source: Vis 2009: 45)

In this particular result, the two different pathways to UR involve WPP and RIGHT, with the connector + representing 'OR'. The asterisk * represents 'AND'. Thus the outcome UR is explained by the presence of WPP or RIGHT, along with the presence of WSE. The use of * or 'AND' also means that WSE is a necessary cause of UR. WSE is part of both causal pathways, and thus is part of 'all' pathways. Now it is explicit why, by our definition of a necessary cause, WSE is a necessary cause. In a simplified list, the two sufficient pathways are:

> WSE and WPP
> Or
> WSE and RIGHT

The special nature of QCA is to notice that it is not necessary to have both WPP and RIGHT, but either of them.

I provide a possible protocol for doing QCA in Appendix Table A6.3. The textbooks on QCA concur on the basic analytical steps. Controversy has occurred around the cutoff level for consistency. Another issue is whether the remainders (those permutations not found in the data) matter when we manipulate the Boolean terms.

In this chapter, the summary and the examples show that QCA and fuzzy-set analysis seek to find triangular patterns of cases, by testing on the basis of a configuration-wise comparisons, Boolean algebra was used to supplement the standard algebra. The Boolean algebraic summary result comprises mainly a set of terms, all linked by 'OR'. These terms each lead to the outcome Y, and each of them represents some configurations where Y is included in X. The overall result can also include any 'necessary' conditions.

One key implication is that statistical methods would wrongly describe as 'error' the diversity of pathways. The distances of a 'point' from a 'line of best fit' will be large if there must be just one 'line of best fit' or one single solution; whereas in QCA we allow for a series of terms, each being an explanatory configuration for Y. A second issue is that error arises within the sufficiency triangle. In QCA such variations within that space get ignored.

In the results by Vis (2009), for instance, the absence of WPP when the pathway is WSE and RIGHT would be seen by a 'line of best fit' regression approach as generating error in the measurement of the WPP $\rightarrow$ Y relation. In my judgement, it is possible to revise the statistical approach to allow for the true or best model, which involves necessary causes separate from sufficient causes and is best represented by fuzzy sets. The only scholars who did this so far were Eliason and Stryker (2009).

Numerous discussion points arise from the fuzzy-set QCA approach. Fig. 6.5 for example takes up quasi-sufficiency. This is the situation when sufficiency is around or below the cutoff level.

Social scientists have debated whether the values X and Y take on a timeless, unchanging level, as suggested by Ragin, or whether in fact the models are restricted to a specific time period. Without panel data, some critics say the QCA method is fundamentally flawed. Ragin argues instead that the method focuses on essences, which are indeed historically contingent but which, by virtue of being structural over a specified period, can be examined as if they were constants. No one says we would predict forward or extrapolate to other societies based on a QCA model. Instead we go onward: we will retroduce after generating the initial model results. Even if it was so that an X caused a Y in the past, for N = 19 with the many sub-elements (farmers, consumers, etc.), it is not necessarily so in the future. Retroduction will tell us about what has happened, and what is now happening, that has created the patterns in these data, exceptions included.

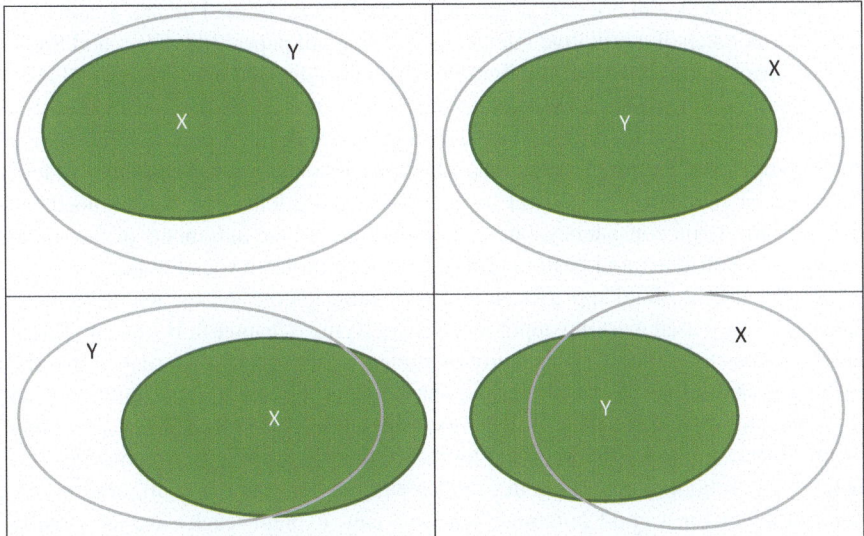

Fig. 6.5 Venn diagrams of sufficiency and quasi-sufficiency. Key: Venn diagrams show cases with X = 1 inside its circle, and X = 0 outside. For fuzzy sets, we imagine the circle being the 0.5 level of the fuzzy-set membership score, which is higher inside the circle and lower for cases outside the circle. Note: Sufficiency is shown in the left column. Necessity is shown in the right column. Quasi-sufficiency is at the bottom left, with the QCA cutoff at 0.80 or 0.75 or similar. Quasi-necessity is at the bottom right, with the QCA cutoff at 0.80 or 0.75 or similar

6.4 Threshold Tests, Quasi-Sufficiency, and Next Steps in QCA

The degree of apparent sufficient causality of a configuration for Y is measured by a ratio known as the inclusion ratio (Smithson and Verkuilen 2006). A cutoff level such as 0.75 or 0.8 is recommended by Ragin (see also Chap. 7). My own research and the F-test based on Eliason and Stryker (2009) also validate the use of such a level as 0.75. However, when sample sizes are small, the F test approach reminds us not to generalise outside the sample even if it is a random sample. The F test aims to consider the distances of points from the diagonal line in fuzzy-set X-Y space that distinguishes whether a case matches, or refutes, the sufficiency hypothesis (ibid.). Sample size does not influence the consistency level as defined by Rihoux and Ragin (2009).

Ragin and Davey (2009: 46) argue for a 0.75 threshold or whatever level of threshold retains a coherent set of configurations above it, and another coherent set below it. Ragin noted repeatedly (Ragin 2000) that often the configurations have consistency levels that vary considerably, and the consistency level threshold can be set in such a way as optimally and conveniently to break up the configurations into those with and without quasi-sufficiency.

Books on QCA aim not only at getting to such a conclusion, but also at assisting in the further step of qualitative interpretation. Seminal papers like Cress and Snow (2000) give lengthy, detailed, ethnographically well-informed interpretation of each configuration. No matter how small or large N is, this method often results in a handful of interpretable (differentiated) configurations of the X factors. The method can also be run on various Y outcomes, which in Cress and Snow's case involve four measures of the success of social movement organisations. The results need not depend entirely on consistency, but also on the interpretive judgments of the social scientists. See Rihoux and Grimm (2006) for applications in social policy areas.

Besides the threshold approach, it is also possible to consider the whole series of steps as a case-based learning approach. Seeing Byrne (Chapter in Byrne and Ragin 2009) discuss the schools in his study of student score success, we realise that the investigation does not stop with finding our which groups meet a threshold. We can dive into the qualitative data of the mixed-methods NVIVO database again and again. Byrne realised, after such explorations, that the use of peer-mentoring was helping one school to achieve better grades without being in the configuration that was sufficient for a good outcome. Thus we can 'explain' each case as it really exists, delving into what really happened. We do not have to take the tabular evidence at face value nor as the only form of evidence.

6.5   Conclusion

In summary, the results of QCA are not primarily numerical. They are primarily qualitative. The Q in QCA means Qualitative not Quantitative. It is a mixed method but more on the qualitative side, yet it uses some numerical data. The researcher can easily mix QCA with other methods from a wide range of options.

Review Questions
Explain in your own words how qualitative comparative analysis (QCA) carries out a data-reduction stage.

Explain the number of permutations of variables for a simple situation with three independent variables and one dependent variable, in which we have data on all the cases (e.g. all the countries of the world, and their population, income-per-capita, longevity, and schooling achievement in years).

Explain what a sufficient cause is when we consider the positive impact of 'social development aid funds' which aim to invest in social programmes.

What does it mean to use a Boolean approach for the outcome 'improved water supply or not improved water supply'. On the surface this appears as a continuous variable, 'amount of water supplied', so what are two reasons why a QCA analyst might choose to represent 'improved water supply' as a crisp set?

(*continued*)

(continued)

Develop an example such as this one, and analyse the necessary causes in it: Suppose the success of bee-keeping in a large allotment or field involves a series of acts, such as buying bees, putting out bee-hive buildings, growing trees, cleaning the hives annually, feeding bees by letting fruit trees flower annually, and planting flowers. Some of these inputs are necessary, some are sufficient, and some are not sufficient in themselves but are a necessary part of a sufficient causal pathway to getting more bees. Make lists of the three types in the grid below, and justify your decisions.

Necessary causes across all cases.	Sufficient cause on its own.	A group of conditions, which together are sufficient, but in which each one is not sufficient.

Further Reading

It is useful to read about causality in Sayer (2000), Lopez and Scott (2000), and Byrne (1998, 2002). A review of the philosophy of science is also helpful (Hughes and Sharrock 1997). The best way to learn QCA is to read excellent applied studies, such as the successful quantitatively driven study by Cooper and Glaesser (2008). Useful reviews of the whole QCA method are found in Ragin (2006), Vatrapu, Mukkamala, Hussain, and Flesch (2015), Byrne (2005), and Olsen (2014).

To study basic themes like consistency, sufficiency, and inclusion, see Caramani (2009).

For help with calibrating fuzzy sets, which is the topic of Chap. 7, a very clear and detailed presentation by Mendel and Korjani is useful (2018). You can find details in Vatrapu et al. (2015), Mendel and Korjani (2012), Smithson and Verkuilen (2006), Longest and Vaisey (2008), Huang (2011), Ragin and Davey (2014), and Thiem and Duşa (2013a, b) of how to use a spreadsheet, macros, or algorithms to do QCA calculations. These sources have much in common, along with minor differences of technique, reflecting how to carry out the Boolean reduction algorithms alongside simple fuzzy-set intersection calculations.

Appendix

Table 6.2 Data for diverse countries measured as crisp and fuzzy sets

id	Country	imfcrisp	invcrisp	procrisp	urbcrisp	ecocrisp	libcrisp	govcrisp	protest	imfpress	urbanise	econhard	invtdept	politlib	govtactv	region	eu	oecd
5	Argentina	1	0	1	1	1	1	1	1	0.83	1	1	0.33	1	1	5	0	0
19	Bolivia	1	0	1	1	1	0	0	0.67	0.83	0.67	1	0.5	0.5	0.33	5	0	0
21	Brazil	1	0	1		1	0	1	0.83	1	1	1	0.5	0.5	0.83	5	0	0
30	Central African Republic	0	0	0	0	0	0	0	0	0.17	0.33	0.17	0	0	0.17	12	0	0
32	Chile	1	0	1	1	1	0	1	0.67	1	1	0.83	0.5	0.33	1	5	0	0
38	Costa Rica	1	1	0	1	0	0	1	0.33	0.83	0.67	0.5	0.83	0	0.83	4	0	0
39	Côte d'Ivoire	1	1	0	0	0	0	1	0.17	1	0.5	0.17	0.83	0	0.67	12	0	0
47	Dominican Republic	1	1	0	1	0	0	0	0.5	0.83	0.67	0.17	0.67	0	0.33	3	0	0
48	Ecuador	1	0	0	1	0	1	1	0.5	0.83	0.67	0.5	0.33	0.67	0.83	5	0	0
49	Egypt	0	0	0	1	0	0	0	0.33	0.33	0.67	0.17	0	0.17	0.5	6	0	0
50	El Salvador	0	0	0	0	0	0	0	0.17	0.17	0.5	0.17	0.33	0.33	0.33	4	0	0
54	Ethiopia	0	0	0	0	0	0	0		0.17	0.17	0	0.17	0.17		12	0	0
62	Ghana	0	1	0	0	1	0	0	0.17	0.5	0.5	0.83	0.67	0	0.5	12	0	0
65	Guatemala	0	0	0	0	0	0	0	0.17	0.17	0.5	0.17	0.5	0.17	0.17	4	0	0
69	Haiti	0	0	0	0	0	0	0	0.5	0.5	0.33	0.17	0.33	0.17	0	3	0	0
70	Honduras	1	1	0	0	0	1	0	0	0.83	0.5	0.17	0.83	0.67	0.33	4	0	0
72	Hungary	0	0	0	1	0	0	1	0.17	0.33	0.83	0	0	0.33	1	1	0	0
74	India	0	0	0	0	0	0	1	0.33	0.5	0.33	0.17	0	0	0.33	10	0	0
81	Jamaica	1	1	1	0	0	0	0	0.5	1	0.5	0.5	1	0.17	0.67	3	0	0
83	Jordan	0	0	0	1	0	0	0	0.17	0.17	0.83	0.17	0.17	0.17	0.17	9	0	0
85	Kenya	1	0	0	0	0	0	0	0	0.67	0.17	0.17	0.33	0	0.5	12	0	0
185	Korea DemocRep	0	0	0	1	0	1	0	0	0.5	0.83	0.17	0	0.83	0.17	7	0	0
179	Liberia	1	1	0	0	0	0	0	0.17	1	0.33	0	1	0	0.33	12	0	0
97	Madagascar	1	0	0	0	0	0	0	0	0.83	0.17	0.5	0.33	0	0.17	12	0	0
105	Mexico	1	0	0	1	1	0	1	0.17	1	1	0.83	0.17	0.17	0.67	2	0	1

ID	Country	1	2	3	4	5	6	7	8	9	10	11	12	13	14
108	Morocco	1	0	0	0	0	0.5	1	0.5	0.67	0.17	0	0.5	6	0
115	Nicaragua	1	0	1	0	0	0	0.83	0.83	0.17	0.5	0.17	0.33	4	0
116	Niger	0	0	0	0	0	0.17	0.5	0.17	0.33	0.33	0.17	0.5	12	0
117	Nigeria	0	1	0	1	1	0.33	0.33	0.33	0.17	0.67	0	0.67	12	0
120	Pakistan	1	0	0	0	0	0.17	0.83	0.33	0	0.67	0.83	0.17	10	0
121	Panama	0	1	1	1	1	1	0.5	0.83	1	1	0	0.67	4	0
124	Peru	1	0	0	1	1	1	1	1	0.33	0.5	0.67	0.67	5	0
125	Philippines	1	0	1	0	0	0.33	0.83	0.5	0.5	0.17	0.83	0.33	11	0
126	Poland	0	0	1	0	1	0.5	0.5	0.83	0.17	0	0.17	1	1	0
129	Romania	1	0	0	1	1	0.33	0.83	0.67	0.17	0	0	1	1	0
138	Senegal	1	1	0	0	0	0	1	0.33	0.17	0.67	0	0.33	12	0
140	Sierra Leone	1	1	1	0	0	0.33	0.83	0.33	0.67	0.67	0.17	0.17	12	0
187	Somalia	0	0	0	0	0	0	0.5	0.33	0.5	0.17	0	0.17	12	0
145	South Africa	0	1	1	1	1	0	0.17	0.67	0.17	0.5	0	0.83	12	0
147	Sri Lanka	0	0	0	0	0	0	0.5	0.33	0.33	0.17	0	0.17	10	0
148	Sudan	1	0	0	0	0	0.5	1	0.33	0.5	0	0.17	0	12	0
156	Thailand	0	0	1	1	1	0	0.33	0.17	0.17	0.17	0.67	0	11	0
157	Togo	1	0	0	0	0	0	0.83	0.33	0.17	0.83	0.17	0.83	12	0
159	Tunisia	0	1	1	0	1	0.33	0.17	0.83	0.17	0.5	0.17	0.83	6	0
160	Turkey	1	0	1	1	1	0.33	1	0.67	0.83	0	0	0.67	1	1
167	Uruguay	0	0	1	1	1	0	0.5	1	0.83	0.17	1	0.83	5	0
170	Venezuela	0	1	0	0	0	0.33	0.33	1	0.17	0.83	0	0.83	5	0
190	Yugoslavia	1	0	1	1	1	0.5	1	0.67	0.67	0	0.17	0.83	1	0
173	Zambia	1	1	1	0	0	0.33	0.83	0.67	0.33	0.83	0	0.5	12	0
174	Zimbabwe	0	0	0	0	0	0	0.17	0.33	0.33	0.5	0	0.5	12	0

Source: Ragin (2000)

Table 6.3 Data for 39 interviewed couples (India, 2006–7)

Y	X1	X2	X3	X4	X5	X6		Configuration
Fuzzy	Fuzzy	Fuzzy	Crisp	Crisp	Fuzzy	Crisp	Number of cases	
0	0	0	1	0	1	0	1	1
0	0	0	1	1	0	0	4	2
0	0	1	0	0	0	1	1	3
0	0	1	1	0	0	0	2	4
0	1	0	0	0	1	1	3	5
0	1	1	0	0	0	0	1	6
0	1	1	0	1	1	0	1	7
0	1	1	1	1	1	1	1	8
1	0	0	0	0	0	1	1	9
1	1	0	0	1	1	1	4	10
1	1	0	1	0	0	0	1	11
1	1	0	1	0	1	1	1	12
1	1	0	1	1	0	1	1	13
1	1	0	1	1	1	1	1	14
1	1	1	0	0	1	0	2	15
1	1	1	0	0	1	1	4	16
1	1	1	0	1	0	0	1	17
1	1	1	0	1	1	1	1	18

Table 6.4 Two protocols for applied qualitative comparative analysis

Protocol Part 1: A methodological plan for hypothesis testing within a retroductive methodology

State the theory or theories.

List the hypothesis or hypotheses.

Collect and analyse the data.

Test some hypotheses.

Do more data collection to explore the situation.

Reflect for a while, and try to explain things:

    a. Why the data show these patterns.

    b. What appears to have caused the outcome of main interest.

    c. Why other scholars would interpret these data differently (discourses).

Frame new hypotheses.

Revisit existing theories and draw conclusions.

Start the next project.

Protocol Part 2: QCA procedures [the stages may be revisited in any order]:

1. Get some of the data organised into cases.

2. Choose an outcome Y and explanatory factors X.

    a. Note that you can work on another Y later.

    b. Note that X can be simply 'associated', not causal, if you wish.[a]

3. Ensure the variates are all either crisp sets taking values 0 or 1, or fuzzy sets.

(IF no fuzzy sets, then it is csQCA. If >=1 fuzzy sets then it is fsQCA.)

Calibrate each fuzzy set carefully using set-theory language. See Röhwer (2011), who discusses the 'modal language', that is the language of ALL or MOST or SOME or NEVER/NONE.

4. Test each main X factor for whether it is a necessary cause of Y.

    a. This test is usually done for the whole set of cases.

(continued)

Table 6.4 (continued)

Protocol Part 2: QCA procedures [the stages may be revisited in any order]:

    b. Sufficiency tests, on the other hand, are for subsets.[a]

5. Test each main Y factor, then sets of 2 or 3, then all of them, for whether each configuration is a sufficient cause of Y.

    a. In this test, you can drop any factor which is 'necessary' for Y.

    b. It is logical to drop them because if they are necessary for Y, the software would include them in all configurations which are sufficient for Y.

6. Now test the X factors for necessary causality for NOT-Y.

7. And test the X configurations for sufficient causality for NOT-Y.

8. [a]If at step 1 you decided not to focus on causality, then in steps 4–7 use the superset idea or the inclusion ratio found in Ragin (2009) and Smithson and Verkuilen (2006), tracing these concepts through into the software. The superset idea corresponds directly to the inclusion ratio and to 'consistency of sufficiency of X for Y' in fsQCA, where Y is a superset of X if all X have Y, but not all Y have the feature X.

9. Decide which is/are the most useful test(s) to finalise.

    a. Re-run the test, and save the interim truth table for re-examination.

10. Decide on whether to omit or ignore some of the cases. Clean the data.

11. Finalise the results, obtaining the parsimonious, the intermediate, and the most complex results, if you wish.

12. Find out the consistency of each important configuration. (If non-causal analysis is being done, then look at the inclusion ratios.)

    a. For crisp sets, also look at the coverage level.

    b. For fuzzy sets, use an X-Y graph to examine the location of cases.

13. Find out the consistency of the whole explanatory set of configurations. (Some call this the model, but I avoid this word 'model' as it sounds universalistic.)

    a. Ragin calls this the solution.

    b. The result may look like this: $A*\sim C + B*D = Z$.

This formula would be read as: Either A and Not-C, or B and D, are sufficient for the occurrence of Z, or for the features of cases to be in the aggregate measured at 0.5 or greater on fuzzy-set Z. The formula can also be read as two pathways to Z. Either the first or the second pathway, or both, appear to be sufficient for Z.

14. Generate suitable user-friendly tables to present your findings.

Note: Hinterleitner, Sager, and Thomann (2016) contain an example of such a result showing some of the technical detail in a supplementary file, available online. This is exemplary in methods terms
[a]It can seem confusing to have two kinds of causality, 'necessary causes' and 'sufficient causes'. For a discussion of the possible overlaps of these types of causes, and a sharp statement of how standard 'frequentist' statistics usually separates causes out into linear additive and separable chunks, see Kent (2009) chapter in Byrne and Ragin (2009). In defining what he means by 'frequentist' statistics, Kent creates almost a caricature of statistical method, and in this book I avoided using the term 'frequentist' because it can cause confusion. The Bayesian school of thought can mean many things, and in Kent's terminology there are three mutually exclusive groups of scholars—frequentist, Bayesian, and QCA (fuzzy set) scholars—but such an overview is oversimplified. Instead, we should think of the world to which we are referring as meriting a variety of methodological treatments depending upon a project's aims, strategies, research questions, and audiences. Therefore we might combine key assumptions from among the three groups of scholars that Kent summarised. For details of what is meant here by necessary cause and sufficient cause, and how to separate them out as variates, please see also Chap. 7 of this book, and Olsen (2019)

References

Bartholomew, David, Martin Knott and Irini Moustaki (2011) *Latent Variable Models and Factor Analysis: A Unified Approach*, 3rd ed., London: Wiley.

Baumgartner, Michael, and Alrik Thiem (2017) "Often Trusted but Never (Properly) Tested: Evaluating Qualitative Comparative Analysis", *Sociological Methods & Research*, XX(X), 1–33. DOI: https://doi.org/10.1177/0049124117701487.

Befani, Barbara, Barnett, C., & Stern, Elliot (2014) "Introduction - Rethinking Impact Evaluation for Development". *Ids Bulletin-Institute of Development Studies, 45*(6), 1–5. https://doi.org/10.1111/1759-5436.12108

Bennett, Andrew, and Jeffrey T. Checkel, eds. (2014) *Process Tracing*, Cambridge: Cambridge University Press.

Bryman, Alan (1996, original 1988) *Quantity and Quality in Social Research*, London: Routledge.

Byrne, D. (2005) "Complexity, Configuration and Cases", *Theory, Culture and Society* 22(10): 95–111.

Byrne, David (1998) *Complexity Theory and the Social Sciences: An Introduction*, London: Routledge.

Byrne, David (2002) *Interpreting Quantitative Data*. London: Sage.

Byrne, David, and Ragin, Charles C., eds. (2009) *The Sage Handbook of Case-Based Methods*, London: Sage Publications.

Caramani, Daniele (2009) *Introduction to the Comparative Method With Boolean Algebra*, Quantitative Applications in the Social Sciences Series, London: Sage. No. 158.

Cooper B. and Glaesser, J. (2008) "How has educational expansion changed the necessary and sufficient conditions for achieving professional, managerial and technical class positions in Britain? A configurational analysis", *Sociological Research Online*. http://www.socresonline.org.uk/13/3/2.html, Accessed June 2013.

Cress, D., and D. Snow (2000) "The Outcome of Homeless Mobilization: the Influence of Organization, Disruption, Political Mediation, and Framing." *American Journal of Sociology* 105(4): 1063–1104. URL: http://www.jstor.org/stable/3003888

Cross, Beth, and Helen Cheyne (2018) "Strength-based approaches: a realist evaluation of implementation in maternity services in Scotland", *Journal of Public Health* (2018) 26:4, 425–436, https://doi.org/10.1007/s10389-017-0882-4.

Danermark, Berth, Mats Ekstrom, Liselotte Jakobsen, and Jan Ch. Karlsson, (2002; 1st published 1997 in Swedish language) *Explaining Society: Critical Realism in the Social Sciences*, London: Routledge.

Dobson, Annette J., and Adrian G. Barnett (2008) *An Introduction to Generalized Linear Models*, 3rd Ed., Chapman & Hall/CRC Texts in Statistical Science, London: Chapman and Hall/CRC.

Eliason, Scott R., & Robin Stryker (2009) "Goodness-of-Fit Tests and Descriptive Measures in Fuzzy-Set Analysis", *Sociological Methods & Research* 38:102–146.

George, Alexander L., and Andrew Bennett (2005) *Case Studies and Theory Development in the Social Sciences*, Cambridge, MA: MIT Press.

Goertz, Gary (2017) *Multimethod Research, Causal Mechanisms, and Case-Studies: An Integrated Approach*, Princeton: Princeton University Press.

Harriss-White, Barbara, Wendy Olsen, Penny Vera-Sanso and V. Suresh. (2013) "Multiple Shocks and Slum Household Economies in South India." *Economy and Society*, 42:3, 398–429.

Hinterleitner, M., Sager, F. and E. Thomann (2016) "The Politics of External Approval: Explaining the IMF's Evaluation of Austerity Programs", *European Journal of Political Research* 55(3): 549–567.

Huang, R. (2011) *Qca3: Yet Another Package for Qualitative Comparative Analysis*. R package version 0.0–4. Pp 88.

Hughes, John, and Wes Sharrock (1997) *The Philosophy of Social Research*, 3rd ed., London: Longman.

Kent, Raymond, Chapter 15 of R. Kent (2007), *Marketing Research: Approaches, Methods and Applications in Europe*, London: Thomson Learning.

Lam, W. Wai Fung and Elinor Ostrom (2010), "Analyzing the dynamic complexity of development interventions: lessons from an irrigation experiment in Nepal, *Policy Science*, 43:1, pp. 1–25. https://doi.org/10.1007/s11077-009-9082-6.

Longest, Kyle C., and Stephen Vaisey (2008). "Fuzzy: A program for Performing Qualitative Comparative Analyses (QCA) in Stata," *Stata Journal*, 8 (1):79–104.

Lopez, Jose J. and Scott, John (2000) *Concepts in the Social Sciences, Social Structure*, Buckingham: Open University Press.

Lucas, Samuel R., and Alisa Szatrowski (2014) Qualitative Comparative Analysis in Critical Perspective, *Sociological Methodology* 44:1–79.

Mendel, Jerry M., and Mohammad M. Korjani (2012) Charles Ragin's Fuzzy Set Qualitative Comparative Analysis (fsQCA) Used For Linguistic Summarizations, *Information Sciences*, 202, 1–23, https://doi.org/10.1016/j.ins.2012.02.039.

Mendel, Jerry M., and Mohammad M. Korjani (2018), A New Method For Calibrating The Fuzzy Sets Used in fsQCA, *Information Sciences*, 468, 155–171. https://doi.org/10.1016/j.ins.2018.07.050

Olsen, Wendy (2009) Non-Nested and Nested Cases in a Socio-Economic Village Study, chapter in D. Byrne and C. Ragin, eds. (2009), *Handbook of Case-Centred Research Methods*, London: Sage.

Olsen, Wendy (2012) *Data Collection: Key Trends and Methods in Social Research*, London: Sage.

Olsen, Wendy (2014) "Comment: The Usefulness of QCA Under Realist Assumptions", *Sociological Methodology*, 44, 10.1177/0081175014542080, July, pgs. 101–107.

Olsen, Wendy (2019) "Bridging to Action Requires Mixed Methods, Not Only Randomised Control Trials", *European Journal of Development Research*, 31:2, 139–162, https://link.springer.com/article/10.1057/s41287-019-00201-x.

Pawson, R (2013) *The Science of Evaluation: A realist manifesto*. London: Sage

Pawson, Ray, and Nick Tilley (1997) *Realistic Evaluation*. Sage: London.

Ragin, C. C. (1987) *The Comparative Method: Moving beyond qualitative and quantitative strategies*. Berkeley; Los Angeles; London: University of California Press.

Ragin, C. C. (2000) *Fuzzy-Set Social Science*. Chicago; London: University of Chicago Press.

Ragin, C., (2009) Reflections on Casing and Case-Oriented Research, pp 522–534 in Byrne, D., and C. Ragin, eds. (2009), *Handbook of Case-Centred Research Methods*, London: Sage.

Ragin, Charles C. (2006) "Set Relations in Social Research: Evaluating Their Consistency and Coverage. *Political Analysis* 14(3):291–310.

Ragin, Charles (2008a) *Redesigning Social Enquiry: Fuzzy Sets and Beyond*, Chicago: University of Chicago Press.

Ragin, Charles, and Sean Davey (2009) *User's Guide to Fuzzy Set / Qualitative Comparative Analysis*, based on Version 2.0, assisted by Sarah Ilene Strand and Claude Rubinson. Dept of Sociology, University of Arizona, mimeo. Copyright 1999–2003, Charles Ragin and Kriss Drass; and Copyright © 2004–2008, Charles Ragin and Sean Davey. URL http://www.u.arizona.edu/~cragin/fsQCA/download/fsQCAManual.pdf, accessed 2020.

Ragin, Charles, and Sean Davey (2014) *fs/QCA [Computer Programme]*, Version [3.0]. Irvine, CA: University of California. The instructions for loading are at: http://www.socsci.uci.edu/~cragin/fsQCA/software.shtml accessed 2020. Users will prefer Version 3.0 or later, and must not load Version 2.0 (held at http://www.u.arizona.edu/~cragin/fsQCA/software.shtml, and dated 2009) unless it is for scholarly reasons of tracing through which distinct formulas and methods of fsQCA gave particular results.

Rihoux, B. (2006) Qualitative Comparative Analysis (QCA) and related systematic comparative methods: recent advances and remaining challenges for social science research. *International Sociology*, 21(5), 679–706.

Rihoux, B., & Ragin, C. C. (2009) *Configurational comparative methods. Qualitative Comparative Analysis (QCA) and Related Techniques* (Series in Applied Social Research Methods). Thousand Oaks and London: Sage.

Rihoux, B., and Gisele de Meur (2009) Crisp-Set Qualitative Comparative Analysis (csQCA), in Benoit Rihoux and C. Ragin, eds., 2009, *Configurational Comparative Methods: QCA and Related Techniques* (Applied Social Research Methods). Thousand Oaks and London: Sage.

Rihoux, B., and M. Grimm, eds. (2006) *Innovative Comparative Methods For Policy Analysis: Beyond the quantitative-qualitative divide.* New York, NY, Springer.

Röhwer, Götz (2011) Qualitative Comparative Analysis: A Discussion of Interpretations, *European Sociological Review* 27:6, 728–740, https://doi.org/10.1093/esr/jcq034

Sayer, Andrew (2000) *Realism and Social Science.* London: Sage.

Schneider, C. Q., & Wagemann, Claudius (2006) Reducing complexity in Qualitative Comparative Analysis (QCA): remote and proximate factors and the consolidation of democracy. *European Journal of Political Research, 45*(5), 751–786.

Schneider, C. Q., & Wagemann, Claudius (2012) *Set-Theoretic Methods for the Social Sciences: A Guide to Qualitative Comparative Analysis.* (Original in German 2007) Cambridge: Cambridge University Press.

Smithson, M. and J. Verkuilen (2006) *Fuzzy Set Theory: Applications in the social sciences.* Thousand Oaks, London: Sage Publications.

Thiem, Alrik, and Duşa, Adrian, (2013a) *Qualitative Comparative Analysis with R: A User's Guide,* London: Springer.

Thiem, Alrik and Adrian Duşa (2013b) QCA: A Package for Qualitative Comparative Analysis, *The R Journal,* Vol. 5/1, June.

Vatrapu, Ravi, Raghava Rao Mukkamala, Abid Hussain, And Benjamin Flesch (2015) Social Set Analysis: A Set Theoretical Approach to Big Data Analytics, *IEEE Access,* April 28, 10.1109/ACCESS.2016.2559584.

Vis, Barbara (2009) Governments and unpopular social policy reform: Biting the bullet or steering clear?, *European Journal of Political Research,* 48: 31–57.

Calibration of Fuzzy Sets, Calibration of Measurement: A Realist Synthesis

7

> **Tip** The case-study method is carried further by fuzzy-set specialists, who see 'variables' as characteristics of interwoven case-study entities (cases). Two key issues pointed out by fuzzy-set specialists are (1) the variables may not be defined or measurable for all cases, yet even non-comparable cases can be compared; (2) the one-way causality sometimes assumed in statistical models assumes that every cluster of cases has the same directional causality, whereas case-study evidence often points towards clusters having distinct causal mechanisms. In this chapter I take QCA further and list several procedures that would enable the reader to do a short-term mixed-methods project. You may want to systematise some data, add to it, look for patterns, discern necessary and sufficient cause, and draw conclusions using mixed logics. Along the way, you may want to probe new, locally collected, or documentary primary data. I provide several illustrations with data examples in the Appendix. Key points are illustrated by interpreting using evidence from a primary field study in India.

The first topic is the measurement of the variables. (It is possible to use the saved factor scores from the confirmatory factor analysis of Chap. 5, and that may play a role in your research design.) The second topic is how to gauge whether the causality is necessary or sufficient. Here there are several interesting terms: 'instances' with positive outcomes; 'permutations' and 'remainders'. These bring us out of the 'random variables' world and into the 'variates' world of primary fieldwork. Remember, the variables approach is for when we use algebra to summarise what might occur in the population, given our sample data. The variates approach is for when we know our cases well (and may or may not have a random, representative, or some other kind of sample). The third section shows how one might begin to gauge how good the fit of the model is in a particular new sense, combining ideas from statistics, Bayesian statistics, and QCA. Here the 'deviations' or distances are

© The Author(s), under exclusive license to Springer Nature Switzerland AG 2022 157
W. Olsen, *Systematic Mixed-Methods Research for Social Scientists*,
https://doi.org/10.1007/978-3-030-93148-3_7

defined and aggregated; results follow closely what we learn from QCA sufficiency tests.

All these guidelines are firmly based on the existing literature. For calibration, we can follow Ragin and Davey (2014) and Thiem and Duşa (2013). The Boolean and fuzzy-set algebra literature is also followed carefully. In this chapter's last part, I followed the ideas of Eliason and Stryker (2009) where a distance function is created that can respond to the information available in QCA tests of sufficient cause. If you are doing non-explanatory research, this chapter can still be helpful. This chapter helps make calibration of fuzzy sets comparable across diverse variates. It also helps to link statistical discourses and the qualitative interpretative approach that is common in applied fuzzy-set QCA.

The innovations in the online annex enable tests of a sufficiency hypothesis. Here the deviations per case are ignored for cases within the sufficiency triangle, and for those which lie outside the sufficiency triangle the deviations are aggregated to make an error measure (also called the distance from sufficiency). This is done for each configuration, not as a universal exercise. As a result of these innovations, sufficient pathways, also called sufficient causes, can be checked for their statistical likelihood of being sufficient for an outcome Y while relaxing the assumption that any cause will operate across the whole set of cases. This allows an open door between QCA pre-processing of datasets and statistical post-processing. Technical details are found in the online annex.

These methods overlap at crucial junctures with statistical methods. They are also well suited to embedding in more complex arguments. In the last paragraphs I therefore discuss how qualitative methods would affect your project. That effectively is where I advocate using scattergrams with selected case-names showing; bar charts or histograms with your key cases identified; and other visualisation tricks to make deep linkages between the more quantitative and more qualitative data.

7.1   Two Forms of Calibration: Ordered Categories or Fuzzy Sets

▶   **Tip** The clusters of cases used in fuzzy-set measurement are called configurations. The measurement metric is the same for all cases, yet heterogeneity of causality is assumed for some groups, that is for relevant sets of permutations. A permutation is a broader concept than a configuration in this sense: not all possible permutations have ever had any instances in the actual society. Therefore, measurement by re-calibrating variables can respond to the concrete actual cases rather than referring merely to the global literature or to theory.

We could summarise the options for calibration of a single condition in a broad table, see Table 7.1. A useful overview of QCA when used in conjunction with statistical testing is provided by Meuer and Rupietta (2017). There are excellent

Table 7.1 Levels of measurement in qualitative comparative analysis (QCA)

Binary	Ordinal categories	Multinomial categories	Continuous variables or variates
Absence = 0 or presence = 1	Set those cases not in the qualitative set as 0, and those 'in' as 1; in between create a series of levels. Avoid 0.50.	Make several crisp sets	Decide on which lower value is going to represent 'not in' and collapse all below that into 0. Collapse the highest values above the upper threshold to 1. Then transform the remaining values, using 0.50 as a crossover point of ambiguity.
Member of set = 1, otherwise 0	Or using SPSS, create cutoffs which break up the distribution well into groups that have roughly equal shares of cases, perhaps 7 or 9 groups. Avoid using 0.50.	Or add these up to make a count, then fuzzify the count	After the upper and lower squeeze, transform using a log function.

Note: The superiority of one transformation over another depends on the actual cases and what is happening, including the possibility that causality works differently according to context

examples of diverse forms of manual calibration and the adaptation of scales to fuzzy-set membership scores in Hinterleitner, Sager and Thomann (2016 including supplementary material), and in Mendel and Korjani (2018). Appendix Table 7.2 shows some raw data prior to fuzzification.

We can choose from these options in a way that responds to the intrinsic nature of each causal condition, or the concept or measure, in real life. See Table 7.1, where categorical, ordinal, and continuous characteristics are subjected to alternate measurement metrics. Appendix Table 7.3 illustrates with an example of a finalised data-table. In fuzzy-set measurement, we can convert any continuous measure to the fuzzy measurement scale 0–1. Ordinal categories have to be forced slightly because the fuzzy set has potentially more precision than the original datum. (Future research may optimise the fuzzification of ordinal measures. Based on the models in Chap. 5 there is hope that improving fit might lead to backward correction of the calibration.) For current research there are three questions to be asked.

1. Is there a natural membership score interpretation? Ragin has stressed this, because the scores should respond to real contrastive conditions.
2. Is there an argument for harmonising a group of variates and making their fuzzy levels equivalent and comparable?
3. Is there a need for choosing the most suitable crossover point where a case will be neither in nor out of the set? The answer to this third query could trump both of the above queries 1–2. Then the researcher returns to looking closely at the upper half of the fuzzy set, above 0.5, and then the lower half, less than 0.5. Asking queries 1 and 2 again, we are then ready to finalise the fuzzy-set definitions.

As shown in Table 7.1, the result is a single fuzzy set. It may have utilised a lot of qualitative information. We can create fuzzy variates from qualitative date in a database, such as a NVivo or Maxqda. As long as the defining points are applied in a coherent way, naming as a number on the original scale 'the fully in point', 'the crossover point', and 'the fully out point', we are able to get on with next steps. These thresholds relate to how sure we are that the case conforms to the definition of this fuzzy set.

An intriguing debate about measurements centres on the idea of a cause having ordinal values, or steps. The software TOSMANA can carry out Multivalued QCA. This free software is also excellent for rectangular Venn diagrams in up to five dimensions based on the truth table. Debates about this clearly place it between the crisp and fuzzy ends of a spectrum. What is not intended is a dose-response function. It is not meant to estimate the slope of response to a dose of medicine for a whole bunch of equivalent, almost identical patients. Indeed, overall mvQCA is not meant to be 'methodologically individualist'. Like the rest of QCA, it leaves scope for holistic causation alongside the Multivalued variates. Or the Multivalued variate if it is a single outcome with multiple observation possibilities. The holistic part of an analysis may be contextual, and the other variates can be causal mechanisms or proxies (see Chap. 8 for illustrations). All in all, the QCA tools have the same freedom to place S I M and E elements into the causal framework as a clear theoretical framework, whether they use one measurement metric or an array of them.

In terms of naming a model, all models that have at least one Multivalued variate are mvQCA. Those with at least one fuzzy set are called fsQCA. The label crisp-set QCA (csQCA) is restricted to situations where every set membership is binary.

I offer examples from Chap. 2 exemplars to illustrate and recall that all the variates are listed in the Appendix.

The chapter first illustrates two forms of calibration—by hand with ordered categories to make a fuzzy set and using calibration in fsQCA to make a fuzzy set.

To create a fuzzy-set membership score by hand, look first at what data you have to hand, and what is its level of measurement. Each variable will be handled separately, and carefully taking notes you will create the fuzzy sets. A binary variable will almost certainly become a crisp set. (A crisp set is also a subtype of fuzzy sets, just taking the extreme set membership values 0 and 1.) An ordinal variable can be calibrated by hand, item by item. You may want to consider the other variates because making all of the calibrations in analogous ways would help with comparability. Examples found in the Appendix show levels of fuzzy-set values; see also Olsen (2009) for nuts-and-bolts exposition.

Figure 7.1 shows how fuzzy sets look when they have been created from raw field data, case by case, with a view to comparability and with an ordinal variable as the underlying basis of information.

As usual, alongside the fuzzy sets we see case identifiers and some multinomials and crisp sets. One cannot use the multinomials in the fuzzy-set analysis so dichotomised versions are placed in three nearby columns (social classes worker, worker with land, and farmer or landlord).

hhid	resistfz	class	worker	workland	farmerll	assets	education	pseudman
1	0	2	0	1	0	0.87	0.17	Venkatram
2	0.87	2	0	1	0	0.5	0.5	Khaleed
3	1	2	0	1	0	0.5	1	Kistappa
4	0.87	2	0	1	0	0.67	0.33	Pullayya
5	0	2	0	1	0	0.33	0.17	Chinnayapɪ
6	0	2	0	1	0	1	0.67	Khaleel
7	1	2	0	1	0	0.5	0.87	Chandran
8	0	2	0	1	0	0.87	0.67	Sridhar
9	0.87	4	0	0	1	0.87	1	Ramaiah
10	0.87	5	0	0	0	1	1	Vasanth Rɪ
11	0.87	2	0	1	0	0.87	0.17	Chitram
12	0.87	3	0	0	1	1	0.17	Narayana
13	0	3	0	0	1	1	0.33	Manju
14	0.87	2	0	1	0	0.17	0	Akbar
15	0	5	0	0	0	0.87	0.67	Syed
16	0	1	1	0	0	0.33	0.87	Jayanth
17	0.87	4	0	0	1	0.87	1	Venkatesw
18	1	2	0	1	0	0.87	0.33	Ranga
19	0	1	1	0	0	0	0.33	Govinda
20	1	2	0	1	0	1	0.33	Pavan

Fig. 7.1 Three fuzzy sets and three crisp sets. Note: The dependent variate is shown at the left (resistance to the employers). There are two fuzzy sets at the right. Three crisp sets, each 0/1 variates, are shown. Finally one multinomial variable for class is shown and that was the source for the three crisp sets. (Source: From Olsen 2009)

The last and most intriguing case is when there is continuous measurement of the X or Y variate. You want it to become a fuzzy set. You may use one of two automated methods, or you may adapt these to create your own mapping. Basically the aim is to make a rank-order preserving transformation of each value, and two authors have given us proposals for doing this.

The most general approach is offered by Thiem and Duşa, who suggest a formula to squeeze up extreme values on both ends of the continuous spectrum, whilst spreading out the rest, around a central point known as the crossover point. The notation they use involves thresholds *tau*, with three key thresholds 'ex' if the case is **excluded** from set S, 'in' if the case is fully **included** in set S, and 'cr' (**crossover point**) if there is the 'maximal ambiguity' about their fuzzy-set membership in S (Thiem and Duşa 2013: 89; cf. Huang 2011). On the left-hand side the formula for the fuzzy-set membership μ_S depends on which zone 'b' of the full number line the case was originally scored at. This re-scaling therefore preserves the rank-order of the cases.

$$\mu_s(b) = \begin{cases} 0 & \text{if } \tau_{ex} \geq b, \\ \dfrac{1}{2}\left(\dfrac{\tau_{ex} - b}{\tau_{ex} - \tau_{cr}}\right)^p & \text{if } \tau_{ex} < b \leq \tau_{cr}, \\ 1 - \dfrac{1}{2}\left(\dfrac{\tau_{in} - b}{\tau_{in} - \tau_{cr}}\right)^q & \text{if } \tau_{cr} < b \leq \tau_{in}, \\ 1 & \text{if } \tau_{in} < b. \end{cases} \tag{7.1}$$

Key: μ_s is the calibrated fuzzy-set membership score. b = zones of the full number line. Tau = the thresholds. Ex = exclusion threshold. In = Inclusion threshold. Cr = crossover point, 0.5 in fuzzy set score terms. P and Q are terms which the user can set.

Thiem and Duşa's formula can be used in a spreadsheet or any software like SPSS or STATA to generate a new variable which is monotonically related to the original but is on a 0–1 scale. The options p and q allow a degree of flexibility as scaling factors.

Another possibility is to use the fsQCA freeware to do the same thing.

The online annex helps explore consistency in more technical detail. This ratio value gives a good measure of how well the data pattern in the aggregate fits the pattern expected in a hypothesis. In words, Ragin defines consistency as follows (a) for crisp sets: 'a straightforward measure of the consistency of set relations using crisp sets: the proportion of cases with a given cause or combination of causes that also display the outcome' (Ragin 2008, page 5 of Chapter 3); (b) for fuzzy sets: 'One straightforward measure of set theoretic consistency using the fuzzy membership scores is simply the sum of the consistent membership scores in a causal condition or combination of causal conditions divided by the sum of all the membership scores in a cause or causal combination' (Ibid.).

In fsQCA freeware, the causal analysis of sufficiency is found in the appropriate menu option, but the causal analysis of necessity is found separately under a separate menu option. It is wise to study necessity first, then sufficiency.

Users should go through the QCA protocol, offered in the previous chapter. In brief:

– Necessity of each cause for the outcome
– Necessity of each cause for the absence of the outcome
– Necessity of the absence of each cause for the outcome
– Sufficiency of each cause and all permutations of combined causal conditions for the outcome
– Sufficiency of each cause and all permutations of combined causal conditions for the absence of the outcome
– Sufficiency of the absence of each cause and all permutations of combined causal conditions involving important absences for causes, in relation to the outcome

In summary, up to here this chapter has shown how to create fuzzy sets, leading to a capacity to do QCA analysis with any variable on any scale. The scales and categories can all be converted into useable QCA variables. The whole process should be transparent. In this stage, transparency and sophistication, including being careful to use precise conceptual framings, are key aspects of being 'scientific' in creating situated new knowledge.

7.2   Features of Multiple Hypothesis Tests Using Fuzzy Sets

Sometimes the negation of causes and the negation of outcomes turn out to be central to QCA. This situation may arise if you have mediation effects that offset a strong basic causal tendency. The negative, or converse, of a variate can be seen as a reversal of a fuzzy set. The fsQCA freeware enables you to compute these negations. With crisp sets we may talk about the absence of the variate. One reason you might want to create the negation of the outcome is if you have a normatively loaded, desirable or undesirable, outcome in the first place. Suppose your outcome is women having a job, with 'Not-Y' being women not having a job. You may decide later that it makes more sense to define the outcome as the converse of Y. You simply subtract the fuzzy set memberships from 1, that is reverse the values on the 0–1 scale. Now you have to either rename the variable or use the menu options in fsQCA freeware. (Usually in software we denote the Boolean or fuzzy-set negation using the squiggle mark, thus ~Y for Not-Y.) Now your model is testing which X's affect ~Y. This will work well: what factors led to women failing to have a job?

In summary, coding the dependent variate in the appropriate direction will enable the researcher to focus on a preferred outcome, and control the discourse around that. But at the mathematical level, the results for ~Y are not symmetric with the QCA results for Y. I described the issue in Olsen (2019a) and Ragin has discussed it in depth (2008).

Readers may want to know if it is worth testing mini-models that omit a lot of the contextual variables. If you test a single mechanism, ignoring its contextual variables, it is possible that omitted variable bias will enter into your results. There are two possibilities at least. One is that you have too much variation in Y because you ignored key background factors that affect Y. Another possibility is that you have too little variation in Y because you ignored key mediating factors or contextual barriers to Y, and without these, the net *ex post* variation in Y against X is small. Thus the original crosstabulation or scattergram can be misleading. This is a point that applies to both QCA and statistical testing.

In such circumstances, it may be better to start with a larger model having 15 or 20 variables, or variates. Then move towards smaller models if some variables are shown to be worth dropping out. You are conducting multiple causality hypothesis tests and not assuming the causes operate independently. Appendix Table 7.4 illustrates how each hypothesis can be seen as a bivariate scattergram of fuzzified cases, whose pattern can be statistically tested. The assumptions are different from those used in frequentist statistics.

7.3   Asymmetry of the Causal Mechanisms? Issues Around Counterfactuals

I would take seriously the role of multiple counterfactuals. Using Boolean algebra it can be shown that QCA is superior, has a wider range of applications to causal questions, and should precede the conducting of statistical studies. By fuzzy-set methods we can discern new things, such as: (i) discerning a necessary cause which ideally we leave out of the statistical study; (ii) discerning which causes are operating in groups not alone, and thus would not be captured when statistical tests for each variable are used; and finally also (iii) discerning that the assumption of linear up/down responsiveness being symmetric may empirically be shown by QCA to be a false assumption.

In case (i) the inclusion of the 'necessary' variable will increase the apparent 'error' of a statistical model, leading to a worse 'fit', all else being equal. Yet this finding is misleading. The inclusion of the necessary variable also raises the parameter count and this worsens measures of goodness of fit. If the 'necessary' variables can be omitted, because they have been shown to be omnipresent whenever the dependent variable is high (or yes).[1] In the online annex I introduce a freeware package that I and John McLoughlin created, called Fuzzy Set Goodness of Fit (*fsgof*). The package calculates consistency values and measures the distance of each case's values from the hypothesised sufficiency triangle in a fuzzy-set space. Version 1 of *fsgof* allows sufficiency tests. Version 2 allows both tests of sufficiency and of necessity. The software is fully open to viewing, being written in Python programming language with the results in a simple spreadsheet format including scatterplots and a textual summary of findings.

In QCA, including a 'necessary' variable will increase both parts of the consistency ratio. This could unsettle the ratio of numerator and denominator, so it is better to leave out the necessary causes when calculating consistency of sufficiency. (See also the online annex, where the methods of calculation are shown in both graphical and spreadsheet formats along with formulas.)

In case (ii), the causes may be operating in groups not alone, as sufficient causal pathways. Suppose it is not only required for Y that we have contextual condition Z but also inputs P and Q. Then P would not pass a correlation test because its association with Y may be low. Similarly for Q and Z. But the combination QZP is sufficient for Y and whenever this combination occurs (or the fuzzy values are high), Y occurs (or the fuzzy set Y is high, i.e. high set membership in the quality represented by Y). This implies weaknesses in the discernment of standard regression methods. There are strengths and weaknesses in the logic of this reasoning. On the positive side, the reasoning helps us see why some explanatory models in statistics do not perform well when used for prediction. On the less positive side, the configurational analysis potentially raises a huge number of combined configurations, creating new challenges which it is recommended to solve using a computer due to the many possible permutations.

---

[1] Conversely, QCA can also show that an X^* should be omitted from regression, due to the tendency that Y will always be absent or low when the necessary variable(s) X^* is (or are) absent or low in value. What is meant here is that a test of necessity is meant to apply across the whole domain of the Y variable, and not just within configurations, and that the test supercedes the need to include X in regression. The claim fits both when X^* is positively associated with Y and when X^* is inversely associated with Y, that is positively associated with Not-Y.

In case (iii), the asymmetry of causal operation can be proven using a fuzzy-set algebra. First the empirical findings are derived using the standard QCA protocol. Then, taking a cause X from among the sufficient causes, if that cause X is absent, we cannot deduce that the Outcome Y will not occur.[2] There could be multiple causes, so even if we did have a workable model that showed one particular measured cause was key in a sufficient pathway to Y, this does not mean it is the only cause (Olsen 2019b). In particular, if we had two or three causal pathways, and we have evidence that none of these has happened, it still does not prove that Y cannot happen. This arises from an absolutely crucial assumption common to all QCA studies: absence of closure. Another word for this assumption is that there are no relevant confounders not included in the model. We do *not* assume this kind of closure (for details of the assumption, see Baumgartner and Thiem 2017).

I should stress that some critics of statistical methods, or more specifically critics of frequentist statistical regressions, have been concerned about the assumption of closure at both the modelling and estimation stages (Chap. 4 in this book). If closure is assumed, then confounders are assumed away. The same problem can be raised with regard to QCA. If closure is not assumed, then instead of a nihilistic view that nothing can be firmly known about causality, proponents of QCA have explained in the works, cited here along with many others, that real causal mechanisms can be known qualitatively in their actual context and not merely (and certainly not only) through quantitative methods. Prevalence of a pairing X-Y is not the only way to establish causality.

Meanwhile, it appears a positive advantage of QCA methods that all the main texts in this area recommend not assuming closure. That is, no algebraic model is postulated to 'explain' or represent a simplified system. Instead the thing represented is an open system. Furthermore, the set of data that we have is not postulated to exhaust what could be possible. Causes are not assumed to work deterministically. Thus, there is a spectrum of opinion about closure.

At one extreme however, some recent critics of QCA have made the assumption of closure a key starting-point for a deductive critique. They claim QCA assumes closure, then show that there is confounding or that confounding variables could cause the QCA estimate without those variable to give wrong solutions. This argument does not work well, because upon any reasonable reading, QCA theorists and practitioners were *not assuming closure*. First we do not assume closure because that would be wrong (ontologically) when studying a complex open system. But secondly, there are practical reasons not to assume closure of the dataset, too. An earlier discussion about closure shows the usefulness of gaining practical knowledge which is not easily encapsulated in the QCA data matrix or statistical matrix (Downward and Mearman 2007). This knowledge also makes the closure

---

[2] It needs to be stressed here that 'X' is a convenient marker which can represent either a single variable or a combination of variables, known as the intersection of these variables. Taking A, B, and C, for example, the configuration BC means B&C or B∩C could be symbolised as X here, and found to be a sufficient cause for Y to occur. The paragraph takes up the possibility that the withdrawal of X, that is the loss of BC, may not have a reverse effect upon Y.

assumption unrealistic, they said. Also the uniqueness of future trajectories of key macro institutions affects current interpretations (Morgan 2009). Morgan's example of the Bank of England is indicative: its actions may be misunderstood, and in the long run they can have new effects not yet evident in current and past economic data. The Bank of England as a macro institution can be misunderstood. The wisdom of not assuming closure encourages researchers to examine these institutions in larger, holistic context (Chaps. 8 and 9). Indeed, if it is possible for unique new historical junctures to occur, then we *never have closure*. Therefore QCA and fsQCA would be superior to statistics if statistical researchers had to assume closure.

The asymmetry issue captures the fact that the reduction in X does not always have the reverse of the effect of an increase in X. Instead it is an empirical question whether it does. Let us assume Y is caused in part by X. Without assuming closure we can nevertheless show that the upward movement of Y, associated with an increase in X, does not imply that we can go backward along the Y line (or the Y curve, or the dose-response function). See also Goertz (2017) who correctly argues we are also looking for triangles, so we don't assume linear relations anyway. He shows that another solution in statistics is to use quantile regression to find the baseline edge of the best triangle that shows sufficiency (above it) of X for Y.

The treatment-effects literature would be seriously weakened in its argumentative strength if it were not true that we have discovered 'Not-X causes Not-Y' (sic) once we have shown that 'X causes Y'. This is a convoluted claim, so let me re-state this point and offer examples. We could show that X causes Y in the sufficiency sense, and yet at that point it still is not necessarily the case that Not-X causes Not-Y. A treatment like a vaccine could have its effect, yet the absence of the vaccine does not imply that the person is not immune, for example. They might have a genetic or prior immunity. Take another example. When we move the 'treatment' metaphor into an economic or political context, we also find that we might be able to remove the 'treatment' in the following sense: by assumption a treatment (such as an incentive payment of $3K to teachers) is not available to teachers, but because teachers' incomes vary so much, some would be deemed to have this treatment, and they could lose the $3K over time. We now need to consider four possible change situations over time: total teacher incomes staying the same at a low rate, staying the same at a high rate, rising by $3K over that time, or falling by $3K over that time. My overall claim in this paragraph is that even if we show that the increase of teacher income by $3K is sufficient for Y to occur, it doesn't mean that the loss of $3K would mean the loss of Y or a reduction in Y. Nor does it mean that the absence of the $3K (income being at a low rate over time) implies the teacher does not achieve Y. Overall, many possibilities are there. The causal mechanisms framework of Deaton and Cartwright (2018) which summarises what was meant in the QCA textbooks is superior to the standard assumptions. By the standard assumptions found in regression studies, if we can show that X is sufficient for Y then we've shown it is 'necessary and sufficient' for Y. It is implied in most regression analyses up to now that the evidence of an association would imply both rightward movement and leftward movement are symmetric. The shift to generalised linear models has not dealt with this issue of asymmetry of causes. The use of panel data is closer to approaching a careful treatment of it. The use of QCA is one way to handle the asymmetry. We then separate the necessary general causes from the various sufficient pathways in the given context. If the context changes fundamentally, then we cannot generalise from a model to the new situation.

What happens in QCA is we ask some new questions. We have to be more clear about 'causes': if we mean sufficient cause, we do not mean the stronger operator, often found elsewhere, known as if-and-only-if (IFF). If-and-only-if is a strong operator that assumes closure. (The symbol = in regression might be read as IFF. It may be wiser to read it differently, and this is an area for further research.) In QCA we have not assumed IFF relations. Thus if your treatment is shown to influence outcome Y upward, you still will not conclude that withdrawing the treatment would bring the outcome downward, unless you have evidence about the effect of Not-X on Y.

One reason treatment-effect studies do assume closure is that the authors found it a good-enough assumption under usual conditions when a human body is being treated by a chemical or an operation. Since human bodies are atomistically similar to each other, and the treatments refer to essential features such as the structure of bones and nerves, the closure assumption seemed harmless in medical studies. The problems arise when there are relevant lifestyle factors (meso differentiation), socio-cultural influences upon genetics (macro differentiation), and reflexive agency of the humans receiving treatment (the body is an open system). In this book I have offered ways to offset, modify, complement, or extend treatment-effect studies so avoid the closure assumption in the project as a whole.

More inclusive studies that have more contextual variates are usually superior. A study with fsQCA first and regression second can also be desirable. Binary measures to reflect a regulatory regime, a policy, membership in a club, and other contextual factors are desirable. The software (fsQCA, *fsgof*, TOSMANA, or others) can be used to quickly check how much an explanation changes when a contextual, institutional, or other meso factor is added.

7.4   How to Make and Illustrate Deep Linkages

You can bring quotations from your case studies into your QCA research. Scholars publish their tables along with the interpretations. They keep their NVivo datasets with all the relevant evidence. You might decide to pinpoint selected cases to illustrate key points (Goertz, 2017). Alternative software for carrying out this work is described by Lewins and Silver (2007).

Three diagrammatic forms to help readers understand the deep linkages are:

- A scattergram with four cases labelled, using pseudonyms, and a page of discussion of each one with quotations.
- A bar chart or histogram with the same four cases labelled, showing where they fit into the variations described in the paper. Discussion names these again.
- A scatterplot of all cases on two key independent variates allows readers to see how they are spread out, and the colour or size of the markers can indicate how their responses vary on the Y outcome.

In all cases, be sure to document your analysis carefully and state which key cases fell into which configurations.

In Conclusion: Pluralism and Encompassing Research Are Superior Methods

There are various ways to do closed retroduction in QCA. The data table itself will be a limiting factor. If you use secondary data, then your QCA can use extra variables or make amended variables such as index scales, but you may be limited to using the available data. Yet, a project could effectively add contextual variables, such as regulatory variables or commercial summaries which vary by locality, using newspapers or official statistics as your data sources. Using your imagination, and matching by locality or region, firm name, or another identifier, you can supplement your original case wise dataset. You may call this administrative data linkage.

If you decide to do open investigation, such as participant observation or moving back into the field for more cases, then you will have an opportunity to code up some new variates or refine the definitions of the variates. Keep in mind the costs involved: If N was 40 and you add 10 new cases, the costs also include revisiting the original 40 cases to make improvements in their data as well as adding new 10 whole case studies A better dataset may result, involving open retroduction, but at a cost. As always sample selection is bound at the upper end by the limits of time and cost.

Review Questions

Take up one of these questions, and answer it, stressing pros and cons: If the QCA fuzzy set allows ordinal measurement, are the results from QCA fuzzy-set analysis analogous to the ordered logit model? OR if the QCA crisp set allows a binary dependent variable, then are the results from QCA analogous to those from logistic regression modelling?

If you have qualitative evidence for only about half of the cases in a study, and you have survey data on all the cases, would you tend to prefer to use QCA or regression analysis? Explain why you give one answer, and possible reasons why you might give the other answer.

Explain what fuzzy sets are. Using an example from the appendix, what is the negation of a fuzzy set?

Can you name the dependent variable in a regression where you would suggest that the direct causes and the helpful explanatory conditions work in groups (as configurations) and not just on their own? If you are unsure, you might choose one from this list of possible dependent variables: years married before divorcing; length of stay in hospital with COVID-19; the number of close friendships someone in a high-school class has.

Further Reading

A superb work on calibrating variates is Mendel and Korjani (2018). Aiming at data scientists, they provide a step-by-step analysis of the methods used in QCA case-study research (see also Mendel and Korjani 2012). For those who are keen on treatment effects and counterfactuals, the article by Befani et al. (2014) covers the bases well. Both these sources promote QCA and providing the reasoning behind calibration and generally the re-transformation of variables.

A few authors have linked regression and QCA; Vis (2012) gives a helpful comparative overview. Quite a few articles produce both QCA and statistical results (Harriss-White et al. 2013; see also www.compasss.org (bibliography), where for each article, the actual data tables and the article text are all held in order to support doing replication studies). Any user could generate Bayesian results for ordinary regression situations using stan or rjags packages in the R software. There are symmetry assumptions in statistical analysis that may not fit some parameters well in a QCA setting. Therefore we cannot use existing statistical software methods, such as regression, to do qualitative comparative analysis (QCA).

A pleasant introduction to Bayesian methods in statistics is Woodward (2012) who uses a simple spreadsheet with macros to carry out Monte Carlo simulations. The essence of a simulation method is that if you simulate 10,000 times from the sample data, you can estimate the distributions of each parameter of a model. To apply this to QCA results would be original, because we showed errors as one-sided under the sufficiency hypothesis. The advantage of innovations that use Bayesian methods is that even for small samples, results giving parameter estimates (with their own distributions) can be obtained, on the basis of an assumption that this sample represents a subset of a hypothetical super-population (Gelman and Hill 2007).

Appendix

Table 7.2 Creation of count data for households in rural India, using interview transcripts (N = 39)

Number of Households	Conform with usual kuulie norms	Avoid kuulie work within village	Exit from village	Resist within kuulie work	Innovate to avoid kuulie	Join in collective action	Special notes
1	C1	A2 coir	A1 son left, 5 sons here				V. Poor
2		A1	E1		I3		
3	C3	A3	E1	R1	I1	J1	
4	C1	A1					No land!
5	C3	A1	E1		I1 vegs	J1	
6	C2	A1	E2	R1			
7	C2	A2		R2	I1 cow	J1	
8			E3		I1 cow	J1	

(continued)

Table 7.2 (continued)

Number of Households	Conform with usual kuulie norms	Avoid kuulie work within village	Exit from village	Resist within kuulie work	Innovate to avoid kuulie	Join in collective action	Special notes
9	* NO.	A1 postmaster				Patriarchal	Act as landlord, don't do casual kuulie work
10	* NO.				I1	J2	Landlord but he got chikungunya fever, a deadly illness
11	C2			R1			
12					I1 silk	J2 JMJ and SHG	
13	* NO.	A1 rents	I1 son left		I1 to get better work		Males work as farmers (ryots)
14	C1	A1		R4			
15			E1				Driver in town
16	C1	A1 rents	E1	R1	I2 cattle, exchange labour	J1 credit group	
17	–	–	–	–	–	–	Landlord
18	C2	A1		R3 resist, make them wait.	I3 sarpanch, ration shop, cow	J3 all chits	Ex-village leader (sarpanch)
19	C1	A1	E2		I1		
20		A3		R1	I1 cow	J2	JMJ as well as DWACRA/Velugu
21		A1			I2 had cattle on lease and tamarind on assigned land		
22	* NO.				I2 tractor, Sarpanch		Wealthy ex-sarpanch of Miniki
23			E2				Seasonal outmigrant, very poor
24	C2	A1 silk, land		R1			
25			E1		I1 tomatoes		
26		A1		R1		J1	
27	C3	A1		R1			
28	C4			R3			
29	C1	A1		R1			
30		A1					
31			E1				
32	C5			R1			

(continued)

Table 7.2 (continued)

Number of Households	Conform with usual kuulie norms	Avoid kuulie work within village	Exit from village	Resist within kuulie work	Innovate to avoid kuulie	Join in collective action	Special notes
33	* NO.		E1 son away at fancy school				Not kuulies. Ryots
34	C5	A1 bulls	E1 land	R3	I3 all land		
35		A1 land contest	E1 daughter working as kuulie away				Very poor
36	C4						
37		A1	E1 sugar factory	R1	I1 EGS		
38		A1 exchange labour	E1	R1	I1 cow	J2	
39	* NO.						Landlord

Note: Hh means household. Kuulie means casual labour. In the cells, the codes A1, E1, J2 mean that we have counted the number of such incidents (1, 1, or 2 instances respectively) as reported in the interview. Codes An measure the number of incidents of avoidance, En counts the number of exits, and so on. Thus the counts shown as, for example, E1 or E2 are one or two exit instances appearing in that case's interview transcript

Source: Field visits in Andhra Pradesh, India, 2006–7. Semi-structured interviews with 1–2 adults, usually the couple who manages each household

Table 7.3 A data block representing fuzzy-set analysis of social policy reform (Vis, 2009)

Government	WPP	WSE	RIGHT	x4null	x5null	x6null	UR_y	y2null	y3null	y4null
Lubbers I	0.33	0.83	1	0	0	0	0.83	0	0	0
Lubbers II	0.17	0.33	1	0	0	0	0.33	0	0	0
Lubbers III	0.33	0.67	0.6	0	0	0	0.67	0	0	0
Kok I	0.17	0.4	0.4	0	0	0	0.67	0	0	0
Kok II	0.33	0.33	0.4	0	0	0	0.17	0	0	0
Balkenende II	0.67	0.67	1	0	0	0	0.83	0	0	0
Kohl I	0.17	0.33	1	0	0	0	0.33	0	0	0
Kohl II	0.33	0.17	1	0	0	0	0.17	0	0	0
Kohl III	0.17	0.33	1	0	0	0	0.33	0	0	0
Kohl IV	0.67	0.67	1	0	0	0	0.67	0	0	0
Schroder I	0.33	0.4	0	0	0	0	0.17	0	0	0
Schroder II	0.83	0.83	0	0	0	0	0.83	0	0	0
Schluter I	0.33	0.33	1	0	0	0	0.33	0	0	0
Schluter II	0.33	0.6	1	0	0	0	0.67	0	0	0
Schluter IV	0.33	0.67	1	0	0	0	0.17	0	0	0
Schluter V	0.6	0.67	1	0	0	0	0.33	0	0	0
Rasmussen I	0.17	0.17	0.4	0	0	0	0.17	0	0	0
Rasmussen II and III	0.6	0.6	0.25	0	0	0	0.83	0	0	0

(continued)

Table 7.3 (continued)

Government	WPP	WSE	RIGHT	x4null	x5null	x6null	UR_y	y2null	y3null	y4null
Rasmussen IV	0.33	0.33	0.25	0	0	0	0.67	0	0	0
Thatcher I	0.17	0.83	1	0	0	0	0.83	0	0	0
Thatcher II	0.33	0.33	1	0	0	0	0.67	0	0	0
Thatcher III	0.33	0.67	1	0	0	0	0.67	0	0	0
Major I	0.33	0.6	1	0	0	0	0.67	0	0	0
Blair I	0.17	0.33	0	0	0	0	0.4	0	0	0
Blair II	0.33	0.33	0	0	0	0	0.33	0	0	0

Key: The names offer prime minister labels for governing coalitions in particular time periods (Column 1). The outcome is UR = Unpopular Reforms. Explanatory variates are:
>   WSE = Weak Socio-Economic Situation
>   WPP = Weak Political Position
>   RIGHT = Rightist Government

Note: This data block contains a few variables and some empty columns. The empty columns help to place the outcome, clearly marked y after its abbreviation UR, in the seventh column, suitable for use in our GitHub programme. See online annex for details of how the GitHub programme could be used to measure the sufficient consistency. The GitHub repository offers free Excel spreadsheets and line-wise programmes to create consistency estimates, see URL http://github.com/WendyOlsen/fsgof, accessed 2020 and works cited in the online annex

Source: Vis (2009)

Table 7.4 The configurations arising from Vis (2012)

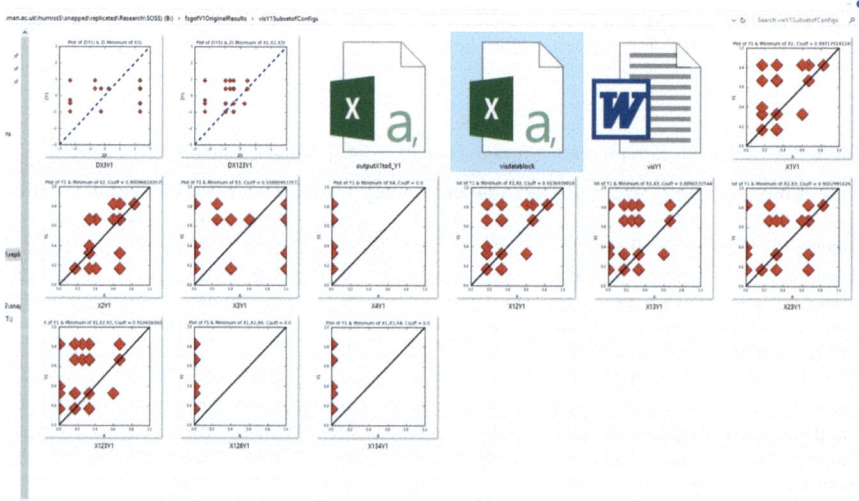

(As seen in the author's freeware, available via URL http://github.com/WendyOlsen/fsgof, accessed 2020 and works cited in the online annex.)

Note: The method of graphing was presented by Olsen and McLoughlin in https://github.com/WendyOlsen/fsgof . Each configuration has been plotted in the fuzzy-set membership score space. At the top left, two examples are provided of the plots for Y by X3 and for Y by X1∩X2∩X3 in the Z score space. Version 1 has a maximum of 4 outcomes and 6 independent variates, while version 2 has more capacity.

The two icons marked X represent sheets of data that hold original data and the consistency results along with Eliason and Stryker (2009) distance results and our F tests for sufficiency. The broad interpretation is that the configurations shown support the following summary interpretation:
>   WSE*(WPP + RIGHT) --> UR (consistency: 0.90)

Table 7.4 (continued)

Source: Vis (2009: 45)

The last ten plots show a scatterplot of Y against every X permutation that occurred

When the online annex is used and our GitHub software is applied, a simple spreadsheet and a Word document are created to hold the results showing consistency of sufficiency for every configuration in the data. The remainders are not presented. The Fuzzy-Set (Goodness of Fit, fsgof) software was updated to allow testing for necessity

We provide all the instructions for users who download the Fuzzy-Set (Goodness of Fit) software from GitHub. It is free

References

Befani, Barbara, Barnett, C., & Stern, Elliot (2014) "Introduction – Rethinking Impact Evaluation for Development". *Ids Bulletin-Institute of Development Studies*, 45(6), 1–5. https://doi.org/10.1111/1759-5436.12108.

Baumgartner, Michael, and Alrik Thiem (2017) "Often Trusted but Never (Properly) Tested: Evaluating Qualitative Comparative Analysis", *Sociological Methods & Research, XX(X)*, 1–33. https://doi.org/10.1177/0049124117701487.

Deaton, Angus, and Nancy Cartwright (2018), "Understanding And Misunderstanding Randomized Controlled Trials", *Social Science & Medicine* 210, 2–21, doi https://doi.org/10.1016/j.socscimed.2017.12.005.

Downward, P. and A. Mearman (2007) "Retroduction as mixed-methods triangulation in economic research: reorienting economics into social science." *Camb. J. Econ.* 31(1): 77–99.

Eliason, Scott R., & Robin Stryker (2009) "Goodness-of-Fit Tests and Descriptive Measures in Fuzzy-Set Analysis", *Sociological Methods & Research* 38:102–146.

Gelman, Andrew, and Jennifer Hill (2007) *Data Analysis Using Regression and Multilevel/Hierarchical Models*, Analytical Methods for Social Research Series, Cambridge: Cambridge University Press.

Goertz, Gary (2017) *Multimethod Research, Causal Mechanisms, and Case-Studies: An Integrated Approach*. Princeton: Princeton University Press.

Harriss-White, Barbara, Wendy Olsen, Penny Vera-Sanso and V. Suresh. (2013) "Multiple Shocks and Slum Household Economies in South India." *Economy and Society,* 42:3, 398–429.

Hinterleitner, M., Sager, F. and E. Thomann (2016) "The Politics of External Approval: Explaining the IMF's Evaluation of Austerity Programs", *European Journal of Political Research* 55(3): 549–567.

Huang, R. (2011) *Qca3: Yet Another Package for Qualitative Comparative Analysis*. R package version 0.0-4. Pp 88.

Lewins, Ann, and Silver, Christina (2007) *Using Software in Qualitative Research Step-By-Step Guide,* London: Sage Publications.

Mendel, Jerry M., and Mohammad M. Korjani (2012) "Charles Ragin's Fuzzy Set Qualitative Comparative Analysis (fsQCA) Used For Linguistic Summarizations", *Information Sciences, 202, 1–23, doi* https://doi.org/10.1016/j.ins.2012.02.039.

Mendel, Jerry M., and Mohammad M. Korjani (2018), "A New Method For Calibrating The Fuzzy Sets Used in fsQCA", *Information Sciences*, 468, 155–171. Doi https://doi.org/10.1016/j.ins.2018.07.050

Meuer, J., & Rupietta, C. (2017). "A Review of Integrated QCA and Statistical Analyses", *Quality & Quantity: International Journal of Methodology*, 51(5), 2063–2083. doi:https://doi.org/10.1007/s11135-016-0397-z

Morgan, Jamie (2009) *Private Equity Finance: Rise and Repercussions*, London: Palgrave.

Olsen, Wendy (2009a) "Non-Nested and Nested Cases in a Socio-Economic Village Study", chapter in D. Byrne and C. Ragin, eds. (2009), *Handbook of Case-Centred Research Methods*, London: Sage.

Olsen, Wendy (2019a) "Social Statistics Using Strategic Structuralism and Pluralism", in *Contemporary Philosophy and Social Science: An Interdisciplinary Dialogue,* edited by Michiru Nagatsu and Attilia Ruzzene. London: Bloomsbury Publishing.

Olsen, Wendy (2019b) "Bridging to Action Requires Mixed Methods, Not Only Randomised Control Trials", *European Journal of Development Research,* 31:2, 139–162, https://link.springer.com/article/10.1057/s41287-019-00201-x.

Ragin, Charles (2008a) *Redesigning Social Enquiry: Fuzzy Sets and Beyond,* Chicago: University of Chicago Press.

Ragin, Charles, and Sean Davey (2014) *fs/QCA [Computer Programme],* Version [3.0]. Irvine, CA: University of California. The instructions for loading are at: http://www.socsci.uci.edu/~cragin/fsQCA/software.shtml accessed 2020. Users will prefer Version 3.0 or later, and must not load Version 2.0 (held at http://www.u.arizona.edu/~cragin/fsQCA/software.shtml, and dated 2009) unless it is for scholarly reasons of tracing through which distinct formulas and methods of fsQCA gave particular results.

Thiem, Alrik, and Duşa, Adrian, (2013a) *Qualitative Comparative Analysis with R: A User's Guide,* London: Springer.

Vis, Barbara (2009) Governments and unpopular social policy reform: Biting the bullet or steering clear?, *European Journal of Political Research,* 48: 31–57.

Vis, Barbara (2012) The Comparative Advantages of fsQCA and Regression Analysis for Moderately Large-N Analyses, *Sociological Methods & Research,* 41(1) 168–198, https://doi.org/10.1177/0049124112442142.

Woodward, Phil (2012) *Bayesian Analysis Made Simple: An Excel GUI* [Graphical User Interface] for WinBUGS, CRC Press, Chapman and Hall.

From Content Analysis to Discourse Analysis: Using Systematic Analysis of Meanings and Discourses

8

Qualitative social-science research methods range widely to include content analysis, grounded theory, thematic analysis, framework analysis, discourse analysis, ethnography, and others. Some methods involve the researcher's participation in a scene, for example participatory ethnography, institutional ethnography, shadowing and action research, and some do not. Participation becomes one extreme of a spectrum which has, at the other end, the use of qualitative data as secondary data. This spectrum runs from high to low engagement of the researchers with the participants. In this chapter I want to link some of the qualitative options with systematic mixed-methods research. I will focus on just part of the qualitative methods terrain in order to suggest, explore, and extend approaches that generate or support a systematic research design.

This chapter is structured from general to specific. There is some broad background first, making the analysis stage distinct from the data-gathering stages; then a close study of one example of a content analysis; and finally the second half of this chapter engages in a more detailed way with how a critical realist would approach discourse analysis within the mixed-methods setting. The last part of this chapter includes original material from primary fieldwork. I will not assume you know about all these methods, but point out readings where commonly used methods are explained at more length. Further readings, listed at the end, also complement what is said here.

The key argument of this chapter is that a realist and systematic approach is feasible, and often we break new ground not just at the 'methods' level but also in analytical terms, thus making novel and important findings possible. The methods displayed here are not just systematic but also 'integrative', meaning that one can do integrated mixed-methods research (Tashakkori and Cresswell 2007; Teddlie and Tashakkori 2009).

© The Author(s), under exclusive license to Springer Nature Switzerland AG 2022   175
W. Olsen, *Systematic Mixed-Methods Research for Social Scientists*,
https://doi.org/10.1007/978-3-030-93148-3_8

8.1   Methods of Qualitative Analysis and Elaboration of Findings

▶   **Tip** There are many qualitative methods. However, thinking through how to interpret the results of a data-gathering exercise has had too little attention. Try to provide for equal 'analysis time' after a given amount of 'data-collection' time. Use a GANTT chart to manage this breakdown of research stages. One implication is that a project probably only needs 1–2 qualitative methods, not 3–4 of them.

Content analysis is a common method in health research, but it is rare in some other discipline areas such as geography. In 2009 I surveyed the range of both qualitative and quantitative methods to see which were consistent with realism and which fit well with critical realism. The result was a summary showing that data-gathering and data-analysis do not exhaust the research interpretation stage (Table 8.1). I have stressed the elaboration stage by using a grey colour, while the white cells contain standard methods.

The use of quantitative methods is shown in the first row, some participatory methods in row two, and the use of archived or secondary data in row three. Case-study research does not fit neatly into any particular row so it appears in multiple places. From this table one could choose to develop a research design using multiple methods.

Innovation can easily take place when we see that we could combine methods to get new insights and that a project can become more insightful by elaborating further the insights from a given set of data. Grounded theory is widely used in business studies, content analysis is common in health research, framework analysis is applied in policy research, and discourse analysis is commonly found in sociology—but there may be good reasons to use a new method in one particular study regardless of the discipline area. Use a method because it suits the research question or the problem that a team wants to focus upon. An essential point is that a research design should use relevant, feasible methods to approach the selected research topic.

8.2   Qualitative Methods, with a Content Analysis Example

▶   **Tip** Using one method on a small dataset can help guide you towards using the same method on a larger dataset. Become aware of which techniques are computerisable and which require human thinking and decisions. In content analysis, the sampling of texts requires human intervention, but the coding and counting of words or phrases can be partly computerised.

This section gives a content-analysis example as a platform from which to raise the broad question of whether we are able to do pure induction or not. It concludes that meta-analysis and broadly analysing holistic social phenomena is commonly

Table 8.1   Data elaboration methods within realist methodological frameworks

Data collection	Data analysis	Writing-Up; interpretation; elaboration
Questionnaires	Induction (as a technique)	Critical social science
Complex sampling and associated Survey Methods	Retroduction about data	Configurational analysis
	Qualitative comparative analysis	Explanatory analysis
Systematic case-study methods		
	Action research	Explanatory critique
Comparative data collection		
	Evaluation	
Historical enquiry	Grounded theory	Critical theorising
Oral history	Realist social statistics	Reframing of hypotheses
Interviewing	Testing hypotheses -About causal mechanisms	Pluralist modelling
Ethnographic research	-About discourses	Re-theorising
Participatory research	Explanatory analysis at multiple levels	Meta-theorising
Gathering texts and translating		
Organising text, video, audio, and other evidence types in spreadsheets and databases	Content analysis	Pluralism of theories
	Critical discourse analysis	Dialogue across geographic space and across layers of stratified societies
NVIVO database construction	Retroduction from data to 'What must exist in order for these data and these patterns to have been observed?'	
Qualitative case-study development		Debates about 'the good' and progress
	Retroduction from future to present interpretations	Methodological pluralism

Key: The first column gives examples of various research methods. The areas shown in white enable the analysis of data to take place, and include the experience of gathering or organising data. The areas shown in grey list some tasks of analysis and meta-analysis in which the social is distinguished from that which is human or personal (Bhaskar 1997), holistic analysis is engaged in Sayer (2000), Flyvbjerg (2011), Danermark et al. (2002), and conclusions integrate past findings with the new evidence
Note: The activities listed above are not intended to be carried out sequentially. This table draws some inspiration from a similar table in Olsen (2010)

what we do, even when the project has carried out a content analysis. The phrases 'open and closed retroduction' (Chaps. 2 and 3) are helpful for the qualitative analysis stage of such research. The same lesson may apply to other forms of qualitative method. Open and closed retroduction can certainly be applied when the aim is to combine quantitative data with qualitative data, or if one carries out multi-stage action research using engagement as well as empirical data (Table 8.2).

A content analysis can form just one part of that broad picture. Content analysis gives a summary of the topics covered, based on computerised coding or a simple

Table 8.2 Making distinctions between the empirical, the actual and the real

	Domain of the real	Domain of the actual	Domain of the empirical
Macro mechanisms	√ (Example 8C)		
Meso-level mechanisms	√ (Example 8B)		
Mechanisms	√ (Example 8A)		
Events	√	√	
Experiences	√	√	√

Notes: This table draws inspiration from Carter and Sealey (2009: 73) but rows 1–2 are original

classification of the parts of one or more transcripts or other data sources. Mäenpää and Vuori provide one concrete example when they run and analyse six focus groups in Finland using this method (2021). The findings show that a descriptive approach was taken. 'The aim was to describe the experiences of nurses in broaching the issue of overweight and obesity at maternity and child health clinics….Informants were nurses working at maternity and child health clinics (N = 28). The qualitative data were collected by [six] focus group interview[s] in spring 2016 and analysed using inductive content analysis' (Mäenpää and Vuori 2021: 3). The authors concluded that 'nurses need more broaching skills, more training, supervising and tools for bringing up overweight and obesity with client families. The continuity of care in primary health care supports bringing up overweight' (ibid.). This conclusion stands out as not resting firmly upon just the basis set up by the six focus groups (81 pages of textual data). Instead, the findings rest jointly on the transcripts plus the authors' prior knowledge of how training and supervision might enable nurses to act differently than they currently do. It was not claimed that the nurses themselves said they should act differently or be trained or supervised differently. Instead the implications of what nurses said in a focus group were developed by the authors. They claimed that they used 'inductive content analysis' (ibid.: 3). The use of inductive analysis for textual data is usually rooted in transcripts and not the authors' personal views nor in a summary of literature.

▶ **Tip** In content analysis, it is useful to consider the logically flawed interpretive method known as verification. A self-verifying research design would broach a hypothesis, gather data suited to approving the hypothesis, conduct an analysis, and then approve the hypothesis. When setting up any qualitative project component, try to achieve something more than self-validating an a priori hypothesis. Alternatively, write up the findings in a less conclusive way to allow for competing views, and make a critical interpretation of the data.

It can be helpful to approach induction from a critical analysis point of view as recommended by Bowell and Kemp in *Critical Thinking: A Concise Guide* (2015, Chaps. 4 and 6). They present induction as a warranted argument in which pieces of information and claims are presented as propositions, labelled P1, P2, and so on. The conclusions of the argument must rest firmly on P1, P2, and so on and not on unspecified outside information, they say. They use pictorial diagrams with P1, P2, and so on in circles, pointing via one-way arrows to the conclusions C1, C2. If C1 underpins C2 then C1 is a premise for C2 and a one-way arrow would show that C2 relies on C1. A sound inductive argument has conclusions that rest upon its premises. If a conclusion has no premises to rely upon, then it cannot be validly derived from the data or other basic assertions. The authors say that a valid conclusion can arise if all the premises are true. Premises include a range of types of thing: data, claims, conceptual definitions, and perhaps a statement of the spatial/temporal context and the limits or boundaries to that context. (This last type of premise helps avoid ethnocentricity. In turn, stating the boundaries of a study also helps avoid the error of overgeneralisation in the conclusions.) Bowell and Kemp (2015) show that using diagrams to simplify an argument can show both where evidence is irrelevant (unused premises P) and where conclusions are unfounded (unlinked conclusions C) (Fig. 8.1).

In the present case, we have P1=nurses' statements during the interviews and P2= the authors' implicit summary of the literatures on broaching subjects and on supervision of nurses. Without making P2 explicit, the argument cannot be well founded or convincing. Yet if we make P2 explicit, the argument is no longer an inductive argument based purely on the focus groups; it is a synthesising paper whose argument rests upon multiple sources. The paper has two rather large leaps of knowledge: the first analytical leap is saying that broaching a sensitive subject might work well for these nurses; and the second one is saying that supervision of nurses could improve their performance in dealing with clients' obesity. These conclusions of the paper's main argument might be true, but they do not rest purely upon the focus group data.

I will make three points about this example to show how a realist or a Systematic Mixed-Methods approach might differ from standard content analysis. First, the content analysis is unable to be conclusive without drawing upon other knowledge resources, such as: a general understanding of meanings in the society; an awareness of the regulatory environment; knowledge of what hierarchies exist; knowing about nurse supervision in this particular case; and details of the social organisations surrounding the data-collection stage, for example hospital or clinic. Knowledge across this breadth of topics is inevitably needed. That makes the use of 'induction' alone a self-limiting, descriptive, and usually narrow approach. The authors here chose to use something more than induction. They built a synthesising warranted argument involving much more than just focus-group evidence. In general, induction is a safe logical method to derive findings, but most authors go further with their interpretations.

Second, the descriptive element in content analysis is somewhat self-limiting in taking 'topics' and not meanings as the coding unit. Published content analyses are often very explicit about coding strategies, which helps readers assess their validity. Some count the appearances of key words, while others count the occurrence of key phrases and topics, using hand-coding, as seen in our exemplar. Yet the descriptive moment itself, when one summarises which topics came up most frequently, is not very satisfying as a conclusion. This self-limiting nature of pure induction is why many other logics are used in social science. Many research teams often use logical steps that are more controversial. In realism the step of 'retroduction' is often used. For nurses broaching a sensitive issue, we might ask extra questions such as those below.

- A closed retroduction analysis question: What made nurses feel hesitant about broaching this topic?
- A closed retroduction analysis question: Were some nurses supervised less closely, and were they the ones that had more trouble with managing a conversation about obesity and diet/lifestyle?
- An open retroduction question: What facts in the survey data available from other sources would tell us about the kinds of clients who are more likely to be obese, and what specific health risks they and their children face? Do these nurses show that they know these facts?
- An open retroduction question: What meanings underlie the responses, such as deep anxieties about the negative response from a client or the tensions that arise in the room with the client? What evidence will we try to get from expert works on the psychology of nursing or the social-science literature on obesity to understand better the emotions that were expressed during the focus groups?

I have stressed that some questions refer directly to the transcripts, and these are the closed questions. Here it is either possible or impossible to answer the questions; and the answer may be very partial if we enumerate which of the 28 public health nurses showed that they felt hesitant, or were less-supervised, or were more-supervised, and so on . The idea of having to count what proportion of nurses answered in a certain way is rooted in an empirical focus, sometimes called 'empiricism', that insists on the micro-nature of textual data. Yet it is not usually recommended to take proportions from a non-random sample. Instead we often use the rankings never/sometimes/most/all to indicate which respondents mentioned which topics. This logic of modes of response was explored by Röhwer (2011). Further elaboration of the findings will rest upon other, or new, evidence.

The open retroduction questions look more towards features of the surrounding society that help us understand the emotions expressed. We can try to understand ignorance or frustration or anxiety or resistance to discussing the obesity issue. These larger (meso) phenomena are difficult to grasp through a series of focus groups but we can try. It is not a question of counting response types, but reaching an understanding by comparing the literature's claims, one's past experiences, and the transcripts at the meso and macro levels.

I would stress that the research task is not just a data-gathering task, such as this: 'the research task was to describe the experiences of public health nurses', but that the research tasks also include generating a deeper-cutting, more encompassing, nuanced description of the problem the nurses face. In this sense the analysis task then is not just induction, but gaining and expressing a deeper understanding of an open system in which change is incipient. The authors definitely took on a role as progressive forces in an open system. If you are going to promote a certain form of training and a new level of supervision then the evidence, ideally, would be put in place to support that position and how it is rooted in the nature of the nurse-client relationship in its local, current context.

Thirdly, the content analysis method has a danger of being limited in scope. One data source led to one content analysis result. It can be meaningful and helpful to look more at the depth ontology and raise one or more of the following as broadened content analysis: what are the social inequality, educational, or asset differences of the speakers? More generally taking a stakeholder view of different patterns of speech content, what do these imply? Interpreting the meaning of phrasings not just in themselves, but also as intentional speech, we offer re-interpretations. All these broadening moments invoke the idea of people as strategic agents who are acting with an intention to influence others when they speak/write. Therefore content analysis is embedded in a social analysis of the context of speech acts. The later parts of this chapter illustrate these interpretive steps by using discourse analysis, which more explicitly links the meanings of speech to the aims and intentions of speakers, and to the social situation in which they are communicating.

In summary, the three points contrasting descriptive content analysis with mixed methods are:

- Content analysis may tend to be limited in its scope and unable to maintain its factual approach, once the implicit premises have been made explicit.
- Moving on from content analysis can occur by realising the rich multiple meanings of each spoken word or phrase. The meanings may differ according to different stakeholders.
- Triangulation with other data sources can enrich the content analysis.

Thus with mixed methods, content analysis can be made more rich at the analysis stage. A study can do further elaboration upon its own initial conclusions. The elaborated account can draw in evidence from other sources: censuses, surveys, expert witness statements, and many kinds of publications. Therefore, content analysis may best sit alongside other techniques and not on its own.

8.3   Three Illustrations Demonstrating Deep Arguments Based on Depth Ontology

The aim of this section is to illustrate with empirical examples how one can use quantitative methods with qualitative data in discourse analysis and make in-depth interpretations.

I give three examples of substantive findings using the warranted-argument approach: Examples 8A, 8B, and 8C. For each, I provide some background and details before drawing a conclusion. These illustrate variants on a holistic theme, or in other words they show some of the implications of a depth ontology. In general, the conclusions could be reached using various types of data, but these three types tend to be arrived at using qualitative data to complement other sources. In this way, this chapter is much more about holistic entities than the other empirical Chaps. 3, 4, 5, 6, and 7.

It is useful to stipulate a meaning for the key term 'discourse analysis' although there are several variants in use today. Discourses are combinations of communicative acts that fit together, with agents making sense of communicative acts through shared knowledge of discourse norms. Discourse analysis is the full interpretation of the origin, nature, and impacts of discursive acts within their social context. To illustrate, in society, generally, we have a neoliberal discourse, Marxist discourse, several feminist discourses, and so on. Focusing concretely on any particular one discourse, there will be images, videos, actions such as gestures, and linkages with other discourses to take into account, apart from texts and documents (Fairclough 1992). Fairclough pointed out that a range of genres must be considered, too, with genres including advertising, music videos, Internet apps, and so on. I will give examples of discourse analysis based mainly on transcripts of translated interviews, but much more can be done using discourse-analysis methods.

Example 8A: Epistemology and Discourse Analysis of a Database with NVivo Coding

While conducting a recent study in northern India, I engaged in re-analysing my own one-to-one interviews carried out in a South Indian village in 2007 (see also Neff, PhD 2009; and full fuzzy-set analysis in Olsen 2009a, chapter in Byrne and Ragin, eds., 2009). Here I summarise the background, research question, and research design for this mixed-methods study. The main point of this exemplar is to show how NVivo coding looks on a small scale before I discuss the implications of scaling up to a set of over 80 interviews.

Example 8A Context and Research Design: In south India, the social-class inequality includes both economic inequality in access to land and vehicles and also social distance, with status asserted through a variety of ritual display behaviours and via cultural capital and investments in education. The research question we set out to examine was how workers varied in their response to the oppressive worker-management practices, common among employers of casual labour, at the village level. Our expectation was that those workers who had a tenancy arrangement would be stronger in their resistance to employer oppression.

The research design included a random-sample questionnaire covering about 12% of each of two villages in 2006/7 along with 39 interviews which I conducted myself. The PhD of Daniel Neff covered diversity across class, gender, caste, and religious-group lines in how well correlated well-being was with being economically well-off (2009).

Exemplar 8A Findings: Nearly half the sample of households were living in dire poverty relative to national standards. In the poor households, women worked intensively with livestock and carried out a wide range of informal activities as well as domestic work each day. The land rental patterns gave tenants from the small-farmer and worker classes access to land, often supplementing small parcels they owned.

Meanwhile the landlords were the key employers, and we had a range of employers and workers, both with and without tenancies, in the sample of 39 for interview. We targeted women for interview, but in many cases the husbands of the adult women also insisted on being included in the interview. This arose from the high status of myself as an interviewer coming from far away, combined with me being a single woman, so that the presence of a man was considered a requirement by most observers in the village scene. (Some detailed codings are seen in Table 8.3 and see the online annex.)

Table 8.3 Codes and nodes in trees in NVivo for South Indian tenants and workers

2006–7 India 39 interviews, selected codes		
Free coding which is inductive		
Free node	No. of interviews	No. of coded segments
Aged people	2	2
Agricultural change	4	12
Agricultural techniques	2	6
Annual waged labourer	2	2
Borrow from self-help group stops landlord	1	1
Costs of tenancy	4	6
Cost-sharing in tenancy	4	8
Court is only for those with property	1	1
Cow loan	1	1
Disputes	3	12
Disputes over land	4	4
Exchange labour	1	3
Fate	1	1
Fear of landlord ejection	1	2
Festivals	1	1
God and prayer	1	1
Housework against paid work	2	2
Illness	2	3
Investment	4	5
Kuulie labour	3	9
Land exhaustion	2	2
Landlord-tenant relationship	4	11
Men power	1	2
Migration	4	8
Output sharing in tenancy	4	10
Self-help groups	4	5
Social mobility	2	2
The landlord will not lower self to do kuulie work	1	1
Unpaid work	6	9
Tree coding which is retroductive and deductive		
Mode of relationship		
Avoid	6	19
Conform	10	36
Exit	8	18
Innovate	5	14
Join	8	17
Resist	7	27

(continued)

Table 8.3   (continued)

2006–7 India 39 interviews, selected codes		
Strategic approaches		
Bargaining	0	0*
Because it is the moral	2	2
Carefully respect landlord	2	3
Institutionalised practices	1	1
Manipulation	0	0*
Secret power	0	0*
Verbs		
Accepts any kuulie work	1	1
Adjust	2	2
Assumes	1	1
Believe him	1	1
Earn bread	1	1
Fear	0	0*
Feel ashamed	1	1
Negotiates with the landlord-tenant	1	1
Never put pressure	1	1
Observe other villagers	2	2

Notes: The source is my primary field data, based on Neff PhD research (Neff), combined with my own interviews carried out in 2006–7 in rural Andhra Pradesh, India. The tree structures shown arose after retroduction had taken place. All coding was recorded manually in NVivo. *These codes were created to test hypotheses arising from the literature

I defined oppressive practices, incidents of conformity with landlord aspirations, and resistance behaviours in an inductive way based on my prior knowledge from the same area 1995–7 and 2005/6, when I carried out surveys and interviews as well as expert discussions over extended periods in both English and Telugu. In general the 2006/7 interviews showed that conformity was instanced by doing what the landlord wanted, against one's own wishes. Resistance on the other hand took several forms. An initial hypothesis was that resistance would take the form of a verbal refusal of an offer of casual daily labour. This was a common event, yet difficult for workers to do, especially if they owed money to the landlord. Since debt bondage was common, exploitation of workers could exceed what would otherwise be imagined to be normal. We explored labour bondage and the social obligation to accept an offer of paid work. This social obligation is a *mezzo-rule*.

While analysing the interviews, further terms had to be fleshed out, since resistance took several forms (Olsen 2009a, in Byrne and Ragin, eds., 2009). Some families would use technical innovation in their cropping pattern or new livestock choices as exit options to get away from doing casual daily work. Thus they could disrupt a long-standing relationship with an employer. Through the interviews, I also found that some people joined in credit/savings groups to avoid the pressure and power of the landlord. Some people were routinely asked to do casual work without any pay—notably women, who were asked to clean up cow droppings or wash an employer's dishes. This irritating situation is why some people wanted to exit the relationship. One group of workers found employment in nearby towns and villages, so by exiting physically they avoided the direct resistance of the daily face-to-face work offer.

I had not realised during the interviews that these patterns meant there are many, varied ways to resist oppression. Social theories in this area stress the clever sabotage of the harassed peasant (Scott 1977) and the gender aspects of debt bondage. I carried out closed retroduction and coding of the interviews to discover and codify the degree of resistance. While coding, I learned more about resistance.

The hypothesis about worker households was predicated on a social class analysis. The meso-level classes were about seven in all (worker, worker with land, farmer, **landlord**, salaried employee, **trader-moneylender**, and **other business owners**). It was crucial to omit three of the classes, shown in bold, from the hypothesis, because they never do casual labour at all. A multi-level social analysis of gender and class was carried out, with sensitivity to caste and religious groups (mainly Muslim and Hindu, the latter being more prevalent).

The coding of free nodes and tree nodes, explained in Gibbs (2002) or Lewins and Silver (2007), allows us to use multiple logics. The free nodes were a way to summarise the inductive findings. The tree nodes were developed after carefully planning how the findings might look as conclusions. The first tree node represents the main findings, written up in Olsen (2009a, chapter in Byrne and Ragin, eds., 2009). The second tree node represents the analysis of strategies, which was written up in a synthesising manner later in the *Asian Journal of Social Sciences* (Olsen 2009b). The third tree node represents part of the discourse analysis of how the workers described their employer relationships. It is rare to see the methods of qualitative coding displayed in detail, because the code lists tend to be voluminous. Here I have carefully shown the empirical application of the multiple logics approach in mixed methods. The survey analysis is held separately but was integrated with deep linkage at the stage of writing each paper.

The coding of discourses helped me with the analysis of resistance against employers. It could help, too, with other questions such as what factors influence the rental rates paid by tenants. As stated by other authors on research design, a set of semi-structured interviews (SSIs) need a clear focus to work well. One set of interviews cannot be used to examine too wide a set of diverse hypotheses. This is probably why the secondary analysis of qualitative SSI data is so rare. In the present case, these interviews were about work and gender and did not cover land rights, tenancy payments, or seasonality.

Coding is first carried out in a freehand way, attaching labels to chunks of text (nodes and references, in the terminology used in NVivo). Next, I found it useful to group the nodes in a tree structure (like a decision tree): General nodes could contain the references which were quotes under a set of from three to six free nodes. 'Copy and paste' techniques were used, as well as 'cut/paste', to mould the database so that it reflected my developing thoughts. Instead of coding discourses, in this particular project I coded instances of reporting a certain type of event.

The ontological moment of realising there were about six types of resistance was felt as a release from tension. I moved away from the original research question (how much do people resist, based on their assets and education?) to a new research question (what kinds of resistance occur in specific contexts of social and economic resources?). The new question was less explanatory and more descriptive; it was based on induction; and it clarified the way an answer could be framed in either tables of percentages or quotations to illustrate each type of resistance. In summary:

- The initial research question was quantitative and had a confirmatory theory behind it.
- The new research question was qualitative and introduced a new typology of resistance.

In terms of our framework of systematic mixed-methods research, the mechanisms among the tenants were as follows: In general, most could creatively amend their work habits, occupation, livestock choices, and even crop choices, to avoid coming into direct conflict with an employer. Non-tenants who had no land had fewer options thus lacked several mechanisms of avoiding oppression. The idea of a low-income worker making sophisticated, complex choices about how to present themselves to the employers in the village would have surprised the elites of the village. Elites had portrayed workers to us researchers as very 'simple'. We did not find workers simple at all, but rather very complex with detailed knowledge of farming, climate, forests, and markets. The degree of sophistication of farming is widely underestimated by outsiders. In this case, the interviews helped me learn that the degree of social shunning of the landowners by the poor, that is by poor people pushing against a higher income group, was quite high in this South Indian scene. The workers felt that if someone did a labour practice that was unethical, they deserved to be avoided. Those employers later found a labour shortage. A nice, fair farmer on the other hand could easily find workers to carry out harvesting or building tasks.

My focus on interviewing women in villages did not lead to original findings in the gender area. Instead I was encouraged to think of each household as an organic unit with great mutuality and shared decision making. I was often told that women locally had a respected voice—particularly those women who do casual labour outside their homes, and those who manage livestock such as cows, goats, or chickens. I was unsure whether this was valid or whether the poor women would agree with that. With just a single interview per respondent, my doubts could not be resolved in this instance. There are strong gender stereotypes about occupational choice and task allocation. These stereotypes are well known but it began to appear that stereotypes are specific to both the social class of the stereotype-holder and the social class of the working woman. Stereotypes often objectify a person, but I was surprised to find how multiple and fluid the stereotypes were in rural India. Sociological books written in India about India sometimes present rural norms as homogenous or even hegemonic, but I found the norms were multiple, changing, and contested.

The mechanisms of choosing and carrying out a strategy of avoiding the landlord's oppressive practices are real, but hard to 'measure'. The semi-structured interviews literally gave me traces, hints, and stories about such strategies (Table 8.4). I later wrote a paper on strategies which drew upon these case studies and my wider knowledge of the credit markets of the area (Olsen 2009b). Again here, 'mechanisms' worked in each household and even at individual level to ensure

Table 8.4 Evidence and logic in the rural Indian resistance study 2006–7

Example 8A	Concrete example of a mechanism	Evidence source
Mechanisms	**Property** influences whether someone can **exit** casual labour by doing silk production	Closed retroduction, going back through the interviews about sericulture and using survey data to correlate land, other assets, and doing paid labour

Note: Bold is used here for mechanisms. An 'exit', which referred to leaving an employment relationship to start up a new relationship, was defined as a mechanism since it generated many other opportunities, and closed off some opportunities

high levels of repayment of debts. Some mechanisms in both instances are subtle, macro social phenomena: the norm of repayment, the norm of obedience, and the norm of conformity. These three norms however are merely mezzo-rules: powerful figureheads but not able to dictate action. Instead, personal strategies and household action patterns met a variety of needs, reacted to a wide range of impulses and reasoning patterns, and were highly diverse. I found agency to be widespread even though the workers' economic property assets were extremely limited.

Example 8B: Rural Women and Work in a South Asian Context

Example 8B Context and Research Design: In 2014–6 I was part of a project leading teams into rural Bangladesh and rural north central India. Our 41 Bangla and 45 Hindi semi-structured interview transcripts from 2016 were complemented by surveys based on random samples in each of 15 villages, carefully chosen over eastern Uttar Pradesh, northwestern Jharkhand, western Bihar, and Bangladesh. These surveys were repeated twice six months apart, and they included attitudes, assets, work patterns, education, and time-use on a single recall day.

The research question was whether the work of women was a cause of well-being or unwellness, defined from within a woman's life as her experiencing well-being or not. Our expectation was that well-off women would get more respect for doing work outside the house, and the human capital theory had led us to think the well-off women would be more likely to work outside the home, too.

Example 8B Findings: Neither of the two hypotheses found support in the data, for two reasons. First women generally faced harsh criticism and suspicion for working outside their own home. Secondly, women of higher caste and social class were likely to be encouraged to stay indoors and do tasks which could be done at home—whether it was domestic work, house-based livestock work, informal work (stitching, cooking), tutoring children, or childcare.

Among the low-income households, women simply had to do paid, casual work outside the home. They expressed frustration and anger about this. Some also felt proud to be doing such work. The class basis of the casual labour market was evident in every single village in all three Indian states and in rural Bangladesh. The meso-level class structures were highly constraining, and so whilst the idea of a housewife who only does childcare and domestic work was a widespread notional vision of a respectable lifestyle, it was always the women from low-income households with little property who faced barriers and therefore could not meet this ideal.

The meso-structural factors were important here. Table 8.5 discerned both from the questionnaire surveys and from the semi-structured interviews. It was crucial for example that only better-off women could own a cow. The cow in turn offered dignity and respect, even though it would mean walking around a village to water and graze the cow. For some respondents a cow represents a sacred or highly valuable object so it was part of the graduated class structure that those households with neither cows nor land were lower in status than those with cows but no land. With highest social status were those families who owned land, who also owned a cow.

Gender stereotypes closely shaped how men and women were sent to do very different sets of tasks: many men did social visiting and ran key occupations that took up most of their work-time, like fishing, arable farming, trading, or construction. On the other hand, women would do a range of domestic chores, child care, washing clothes, arable farming, 'help' with fishery work or livestock, and sometimes paid casual daily labour. The latter was the type of work most widely considered degrading or risky for women. The gain in well-being that came from being paid and having food to eat was offset by both the drudgery of the casual labour and the social criticism the women potentially faced for not being at home.

Table 8.5 summarises that these meso-level mechanisms (social criticism, drudgery, class comparisons, and class relations) operate in a sphere at a more aggregate and corporate level, compared with the actual influence of assets like land and animals on people's work choices.

I found that shaming of other people was commonly used as an attempt situation in the Indian and Bangladesh rural interviews in 2014–6 (written up with specific reference to India in Olsen, 2020). A social and historical deconstruction of how the society construes each activity can be useful. Here, closed retroduction is not enough. We also can go beyond the villages to read ethnographies and expert accounts of the long-term genealogy of social class in the area (Jeffrey and Jeffrey 1996, for Uttar Pradesh, for example). Here we find that women's position and the contradictory pressures they face have a long and contested history, which finds expression not just in books but in other genres such as poems and songs, drama, and life-histories. Class, caste, and gender all interweave. The use of these materials to help interrogate the social meaning of each phrase is a step beyond the direct,

Table 8.5 Evidence of class and shaming meso mechanisms in the rural India-Bangladesh study of gender and work 2014–6

Example 8B	Concrete example	Evidence source
Meso-level mechanisms	**Class** influences shaming behaviours and the reception of criticism. The class structure has hierarchical aspects. People are aware of the shaming acts	Open retroduction, going back through the codes and nodes on 'criticism' in the interviews, then comparing these with ethnographic write-ups about social class in this region

Note: Class is the meso-level entity here, shown in bold

reflexive interpretation of interviews. We move into the analysis of social trends and must consider macro, cultural, and meso formations.

In analysing the interviews, we had a large corpus (body) of data over a huge geographic area including both rural north and southwest Bangladesh. In both the Bangladesh and Indian contexts, this class-gender contradiction stood out. Here I will pick up on another meso and macro aspect of the issue.

We examined the 86 interviews, which after transcription comprised 840 single-spaced pages. Some team members grasped the interviews in their own language, typically Hindi in India and Bangla in Bangladesh, then English was used to bring the comparative material together. The work was funded by the UK government via its Economic and Social Research Council. The grant title was 'Gender Norms, Labour Supply and Poverty Reduction in Comparative Context: Evidence from Rural India and Bangladesh'. Funding was under grant number ES/L005646/1, and research oversight was carried out partly by the Department for International Development (DFID).

Women's well-being was being undermined by social criticism of key behaviours that low-income women would do. The reiteration of social class superiority was carried out through comparing women of low class with the contrasting high-class female behaviours of being a housewife, having a high earning husband, and being able to raise children without having to do outside work. Both 'low'- and 'high'-class women agreed on a ranking of class positions such that the low-income women deem themselves not doing well, while the high-income women experience more well-being across a range of domains day-to-day. The negative affect of being in a low-income group was, however, offset by certain positive feedback loops for casual worker women:

1. Workers get cash and can pay for food or clothing and other necessities.
2. Women workers felt proud at this contribution.
3. Respect was gained over time by having an occupation and being good at the work.
4. 15% of women had absent husbands, who had migrated away or who had passed away, and these women felt their ability to sustain the household was something to be pleased about.

Thus the resulting condition of the low-income women was very mixed. Tensions were shown, and self-criticism was often followed by words of praise attributing respect to self. Men who got involved in the interviews also routinely complimented their womenfolk on their economic contribution through work.

This analysis has been guided by a keyness analysis to enable the key discourse patterns to stand out. We found the key words shown in Table 8.6 most prominent, relative to ordinary language (details of the extraction method and keyness concept are in the online annex). The analysis of these words led me to distinguish two main discourses in the core area of the research question: a dominant discourse of the masterful breadwinner; and a marginal and yet very common discourse of the woman worker who is making decisions without a male decision maker (Table 8.7).

Table 8.6 Key terms in 39 South Indian 2006–7 interviews, sorted by keyness

Row; Word; Keyness		
166. Mahesh 10.71462	195. Repay 16.26174	225. Postmaster 48.23557
167. Untie 10.71462	196. Agriculture 21.33045	226. Rasool 48.23557
168. Wishers 10.72171	197. Drought 23.26969	227. Wetland 48.81183
169. Village 11.01221	198. Quarrel 23.3069	228. Unpaid 51.75975
170. Wells 11.04629	199. Fieldwork 24.11724	229. Paddy 53.4035
171. Scarcity 11.15269	200. Remuneration 24.11724	230. Tamarind 56.35589
172. Milking 12.0702	201. Irrigation 24.19348	231. Tiffin 64.33759
173. Bengal 12.86635	202. Mango 25.31756	232. Bullock 64.80383
174. Firewood 12.88372	203. Entrusted 26.60123	233. Harvest 71.8233
175. Rains 13.61349	204. Crops 31.41488	234. Krishna 72.51061
176. Irregular 14.27698	205. Chits 32.14531	235. Rabbi 74.27854
177. Tills 14.31557	206. 40000 32.14531	236. Watchmen 80.45139
178. Kolar 16.06679	207. 20000 32.14531	237. Alias 80.74285
179. Satish 16.06679	208. Ioran 32.14531	238. Acres 85.19212
180. Sheik 16.06679	209. 15000 32.14531	239. Manure 90.86118
181. Granary 16.06679	210. Pradesh 32.16807	240. Crusher 96.57697
182. Basher 16.06679	211. Landless 32.16807	241. Pesticides 96.78711
183. Workout 16.06679	212. 10000 32.23654	242. Landlord 102.5266
184. 350ft 16.06679	213. Anymore 32.35134	243. Fodder 104.9579
185. Bangles 16.06679	214. Rental 34.65702	244. Bangalore 105.8908
186. Handover 16.06679	215. Cents 35.47319	245. Rents 118.3166
187. Beetroot 16.07229	216. Disagreement 35.50066	246. Rupees 200.7861
188. Josef 16.07229	217. 50000 40.22478	247. Tenancy 289.3245
189. Hardworking 16.07229	218. Kerosene 40.22478	248. Kumar 291.0077
190. Punjab 16.07229	219. Uncultivated 40.22478	249. Approx 372.5272
191. Behavior 16.07229	220. 30000 40.22478	250. Reddy 2298.322
192. Upland 16.0833	221. Fertilizers 42.97025	
193. Rainy 16.09433	222. Wendy 47.86543	
194. Sunflower 16.10538	223. Tamil 48.23557	
	224. Allah 48.23557	

KEY: The row number rises as keyness rises, and we left out words that had low keyness
The keyness ratio is the odds ratio of the word, dividing its prevalence in the data by its prevalence in the baby British National Corpus. This method is described in the online annex
Notes: This table shows part of a list of the words over four letters long in 39 interviews, after lemmatisation and translation from Telugu to English. The raw word counts are first compared with the overall word length of the interviews. Then this ratio is compared with the relative prevalence of each word in the overall 'baby' British National Corpus (BBNC). The odds ratio is a measure that harmonises the analysis of the words' relative prevalence

Table 8.7 Key terms in one discourse topic, sorted by dominant and marginal discourses

Topic	Money	Family	Work
	Dominant discourse: should repay loans.	Dominant discourse: how one should behave.	Dominant discourse: masterful breadwinner makes decisions.
	Marginal discourse: join a self-help group to borrow to repay loans.	Marginal discourse: deviants did what they needed to do to gain respect.	Marginal discourse: Women forced to work by circumstances are heroic.
Key words	Bribe, bribery Creditors Indebted, etc.	Bachelor Children Dowry	Labour Sewing Workplace

Note: Only selected discourses and selected key words are treated here. There were twelve topics in the data, and each had evidence of from two to four main discourses in these data. Further details are presented in the online annex where keyness calculations are described. The degree of keyness for the nine words shown here was high. In total, 125 words exceeded the keyness cutoff level and at the same time fit well into discourse topics. Another 130 key words had high keyness but did not fit into an interpretive topic. These were words like 'doing', 'district', 'thousand', and 'valuable'. These were generic terms

These two discourses feature some iconic figures in common, for example husband and wife; mother-in-law; tasks and payments. Their differences are profound though. The masterful breadwinner discourse shows men in a good light for supporting their womenfolk and children, and is critical of men who fail by disappearing (they may say he 'deserted her'). On the other side the less common discourse is around women working for pay. Paid work in rural areas of India is sometimes presented as an independent, positive activity of women. It is presented as beneficial as she can then feed her children (thus supporting the mothering discourse but opposing the male breadwinner key to them). Yet at the same time, we found women currently also presenting others' critiques of women being outdoors. Women who work outside the house were aware of criticisms of women being alone in the day-time outdoors, women being at risk of harm and at risk of social censure, and women as recipients of male instructions yet able to do what they judge best. The absent male is implicitly present in both discourses.

Assumptions of the above analysis include: the interviews are balanced in coverage of men and women; the interviews give a clear gaze into the working world; the interview is not biased by a language decision or by translation problems; the generality of the summary is meant to be only to those interviewed, not to the wider population. Thus it is a concretely grounded summary, not a population generalisation. The interpretation of the interview is seen as a non-theoretical, grounded, contrastive summary which pits one discourse (and its patterns and iconic agents) against another one.

In analysing the data on women and work in rural India and Bangladesh, two other pre-existing theories had little to offer. One theory focuses on social exclusion, for example exclusion from the labour market. Here the theory is factually wrong in saying women rarely 'work'. By restricting the meaning of 'work' to commercial work, social-exclusion approaches tend to reinforce breadwinner stereotypes in which only 'men's work' is accepted as valid 'work'. Yet, at the same time, the social-exclusion theory aims for women to be 'included' in commercial work. This theory aims for social inclusion; it aims in part to share and voice the positive affect attaching to paid work for women. 'Affect' here refers to the overtones of positive approval attached to commercial work. It is an emotional dimension that is noticeable in the theory. The reasons for agreeing with this affect would differ from the reasons (or evidence base) that one would use to assess whether women work, in the broader sense of doing concerted activities that constitute neither leisure nor rest. The social exclusion theory was not useful in explaining the data, and instead led to reinforcement and reiteration of the idea that women are unable to gain access to breadwinner work roles.

The second theory, human capital theory, states that more education and skills along with more work experience are associated with higher wages. (There are variants on this theory; some are about investments in schooling, and others are about long-term labour-market outcomes.) One usually says the human capital is higher among higher waged people. To bid for higher wages, one may invest in education; and the payoff expected can have a causal effect on whether men, or women, get educational investments depending on the gender wage distribution. In human capital theory it was pointed out that a rational decision maker in a unitary household might withdraw schooling from girls and invest more in boys, given market conditions that favour boys with higher rewards to formal education and credentials.

Human capital theory did not apply easily in villages due to the prevalence of manual labour, widespread skilled farming work for which schooling did not prepare people, and the highly informal nature of most work. Again, whilst having sympathy with the theory, it does not explain much in this rural situation.

Example 8C: Discourses of Sri Lankan Banking Using NVivo Coding Across Genres

Example 8C, Context and Research Design: I have carried out detailed studies of micro-finance in localised rural areas: once in highland Sri Lanka, using interviews and a random-sample-based questionnaire survey; and again in southern India's Andhra Pradesh, again with interviews and a survey. By 1995, the Sri Lankan government had begun to commercialise a large-scale public-sector banking system (Olsen 2006). Women and men had engaged with this public-sector bank branches in many rural areas, but a rapid expansion through credit cards, house mortgages, and urban lending in semi-privatised banks was planned. The research question we posed was what factors would inhibit or encourage women to engage with micro-finance in rural Sri Lanka. I was contracted as a consultant for the Department for International Development (DFID, the UK aid ministry) to gather both the large-scale consumer finance survey data and some primary field research data in order to shed light on the issues around financial expansion. We conducted a random-sample survey questionnaire and semi-structured interviews with randomly chosen village women in upland Sri Lanka near Kandy. We were confident that women were partly blocked by cultural norms from using banks for savings and loans. The hypothesis was that women might be willing to borrow money in safe, group-secured peer lending schemes.

Example 8C, Findings: We found that savings were highly prized by most women. Debt was widely shunned by the women. A widespread perception was that Buddhist principles inhibit taking debt, and promote saving in advance of major expenditures. We interviewed both Buddhists and Muslims in the area. After translation, the two religious groups could be compared. Buddhist women were more likely to refer to Buddhist principles while Muslim women pointed out that idea of *riba* (interest) was a banned practice based on key Muslim writings. Class and gender differences in debt- and savings-attitudes were muted in this one Sri Lankan village.

In southern Sri Lanka I also gathered bank application forms and interviewed bank managers. The forms added another 'genre' of data to the database. By looking at how bank managers viewed their customers' savings account applications and loan applications, I was able to link aspects of household structure and cultural norms with bank operational rules. I found advertisements in the local newspapers also treated microfinance in a patriarchal way. Both in newspapers and in banking application forms, women were treated only as mothers and daughters, not as active customers in their own right. Women appeared in adverts and on passbooks and annual reports of banks associated with children and cooking. By contrast men in adverts were depicted with tools and men were usually shown at work. The patriarchal aspect of the application forms was this contrast, which was quite general at the time:

- women fill in the form and sign it but must get their father or husband to also sign it.
- men fill in the form and do not need a family member to sign the form.

The Buddhist dominant cultural grouping in Sri Lanka routinely attached assets to men so that the male line would inherit assets whether by dowry or by inheritance. Part of the coding logic is shown in Table 8.8. Banks acceded to this gender-specific norm by ensuring that women could not gather or offer as surety any financial assets without their menfolk signing off their approval.

In Sri Lanka I also explored the Muslim cultural group whose beliefs about the charging of interest were typically very negative. A Muslim norm here is that charging interest is sinful while an alternative method of charging (taking a one-off loan fee or providing a bonus on savings) is widely approved at the normative level. The Muslim cultural minority, which overall formed 7% of Sri Lanka's population at the time, was not well represented among bank users. The rural Buddhist ethnicity people were more commonly using the banks in the study area in rural central Sri Lanka. A 'women's window' was offered by some bank branches to allow women to visit without risk of touching men outside their family. This was a progressive inter-cultural policy gesture. However inside banks, symbols of the Buddhist forms of worship were very widespread, notably lamps and brass artefacts (Table 8.9). (We had already found that in Andhra Pradesh, India, bank staff put up posters of

Table 8.8 Evidence and logic in the Sri Lankan study of microfinance and banking

Example 8C	Concrete examples	Evidence
Macro mechanisms	**Banking culture** is affected by **religious prudence norms** Repayment patterns and savings are affected by **prudence**	Closed retroduction, synthesis, deep linkage: Semi-structured interviews, expert bank manager interviews, local questionnaire, and a large-scale survey on credit and savings

Note: Bold is used for macro entities here

Table 8.9 The key role played by 'Abstract Premises' in a solid argument

Example 8C	Assertions which are Abstract	Evidence
Macro mechanism 1	First, the assertion that culture exists and affects how banks work	We have coded both the photos and bank books, and the text of interviews, about cultural artefacts from Buddhist religious tradition in the Sri Lankan bank offices
Macro abstract assertion 2	Second, having a 'women's window' i.e. bank entrance, for the Muslim women	The concession to a Muslim tradition of female seclusion was notable as Muslim was a local minority group. Thus, 'culture' does not refer only to the dominant culture
Macro mechanism 3	Third, the assertion that social class affects how bank customers act	We coded the agency of the customers, using village interview data. We traced cases through back to their assets, jobs and household wealth

Hindu gods or saints near their desks, and/or they light incense to carry out Hindu worship—practices done in public which Muslim religious people find quite unsupportable.) Overall the Sri Lanka interviews gave people a chance to reflect on their norms and religious views and how these were highly relevant to decisions about banking.

The implications for microfinance were numerous. We used a survey to explore how people differed in their attitudes about savings (e.g. degree of prudence), repayments, borrowing, and joining microfinance schemes.

Overall the Sri Lankan study involved a team of ten people for just two weeks of fieldwork, carried out in three languages. Reflecting on the data, making notes, comparing cases, extracting quotations, and drawing up general points about subgroups were valuable and gave insights that could not be gained from statistical regression. Fairclough's (2001) book on *Language And Power* offered a helpful approach that deals with the complex multiple sets of field materials. Fairclough had accurately set out the notion of 'genre', which discourse analysis could carry forward into the study of meaning, power, and social norms (1992).

I could code the interviews so that the sense-making at the level of language, phases, topics chosen, and how people approach money issues could appear in a coding 'tree', while a closer analysis of language styles and patterns, metaphors and idioms, and power in discourse context could be put into another 'tree'. These coding 'trees' (nested sets of codes) are the collecting nodes to which we copied/pasted the original 'free nodes' that seemed to be the most important aspects of the interviews upon our reflective read-through long after the fieldwork took place.

An online guide to coding uses this example to illustrate trees, free nodes, models, and discourse coding in the Sri Lankan translated interview case (see online annex).

A small research literature takes up whether all premises in a scholarly argument should be true (Bowell and Kemp 2015: chapter 4). My exemplar from South India can be used to illustrate that we do try to have true premises. In this sense, the proposed 'keyness' and 'keyness with discourse analysis' methods have much in common with grounded theory (as advocated by Charmaz 2014).

In small workshops and in a medium-size project, I have used the keyness index and then discourse analysis to look at women in couples discussing their work practices. In this section, I refer only to a single interview which corresponds to the workshop, thus giving a small database for discussion purposes.

True premises can be used to achieve a firmly grounded, realist argument. 'Abstract' premises could be true but not general, while general premises might be unfounded or untrue. These two types of premises are quite different. The inductive moments found in the use of discourse analysis from textual data are fundamentally based on abstract premises, such as a definition of discourse and a statement of how one discourse is working and what it means to people. These premises however are argued (by realists) to be true, and not merely to be subjectively preferred nor just a personal choice. Thus, we invoke a notion of true premises during the inductive logic of carrying out a discourse analysis. The realist 'take' on this is that the notion of true premises is actually true, in the sense that those discourses actually exist. This does not occlude that other approaches which discern other discourses would be false: these could co-exist and be true, too. By contrast, a strong social constructivist 'take' on this could be that the premises need not be true, because they are just assertions as part of a deconstruction of what was said. Whilst this may be arguable, it is not the same as claiming that the very first part of the argument is actually true. Both schools of thought offer rich situated knowledge and my aim here was to show clearly how they differ, without giving a full account of the realist or strong social-constructivist positions. This limitation is due to lack of space.

Many variants on this debate have been put forward. One key point is that there is transitivity of knowledge about qualitative data. Transitivity refers to interactions of self and that which is known. (Intransitive knowledge would be where our knowing about something does not change or influence the existence nor the nature of that thing.) Knowledge of discourses is usually transitive because our own discourses affect what we tend to say about others' discourses. Knowledge in this realm is subject to dialogues going on around the researchers and this is why 'hermeneutics' is needed, not just a single 'interpretation'.

8.4   Conclusion

I used examples from South and north central India, rural Bangladesh, and Sri Lanka to illustrate the analysis of discourses in a mixed-methods context. These examples contrasted with the content analysis example in offering a deeper approach to structural inequality (notably class, gender, and ethnic aspects of the setting). For systematic mixed-methods research, using a wider lens that allows social inequality or the legal and regulatory situation or other broader macro issues to be raised is likely to be feasible. Deep linkage is one label that helps indicate the tracing of processes, ideas, or linked and intersecting causal mechanisms through various data sources. I gave examples to show what the terms 'micro', 'meso', and 'macro' meant in primary field research. I also illustrated the mixed-methods integration of case-studies with the qualitative interpretation of meanings in data. It turned out that stakeholders were numerous, and one voice did not give an adequate description of a scene but did contribute in a meaningful way to a balanced understanding of that scene. The examples tended to appear very complex at the analysis stage, whilst the interview transcripts themselves were manageable and helped to simplify the data.

For those who want to use mixed methods with action research, it may be wise to keep the number of interviews and focus groups low, and/or limit the survey sample sizes, to enable an extensive period of reflection and analysis near the end of a project. That is why earlier in this book I urged that mixed-methods action research might use secondary survey data or secondary qualitative data to make a project manageable. In the annex to this chapter I offered practical techniques for reducing the size of a qualitative textual database in three ways. First one can count and rank the words or phrases; second one can group them into discourse topics and reduce the number of topics covered in one analytical event (e.g. for one conference paper); and thirdly one can look at deviant and standard discourse examples from within the corpus so that summaries of 'prevalence' are less important and exhaustive tables can be avoided. The solutions offered here did not cover all possibilities, and future research will be able to guide us on using video and audio evidence, walk-through recordings, continuous data, and multiple-sensory recording divide data—such as clickers or body monitors—all of which tend to create large datasets. The use of the concept of keyness from linguistics illustrated how a meta-analysis could engage with all the data while being a manageable stage of research that invited analytical elaboration.

Examples 8A, 8B, and 8C illustrated the use of coding nodes, free nodes, and trees. The software I used was NVivo but it is equally feasible to apply R or other software to gain the same outcomes. The discovery of new findings by reviewing and interpreting the data, usually called hermeneutics, can go as far as discovering real mechanisms. These mechanisms may be factors which cause social change; features of social reproduction of the society; or other mechanisms. Some operate at a micro level, and others operate between meso units or from a meso unit (e.g. a firm as a whole entity) upon elements within that unit (e.g. workers). To discern mechanisms that operate upon people, and which affect people even if they are not aware of it, goes beyond ordinary hermeneutics in the following sense. Instead of drawing upon only what was said and what it 'means'—either what the intended or social meanings are of these utterances—we also drew conclusions about what mechanisms were at work. This is the realist step of deriving a knowledge of social institutions, social structures, articulation and contradictions of the structures, and other meso and macro mechanisms, from qualitative data. In this step, the use of survey data proved helpful in all three cases. It was useful for class differences, identifying caste and religious background information, attributing demographic dimensions to people of each gender and age-group, and for drawing contrasts.

Lastly I also showed that macro entities such as the neoliberal discourse, the state, or other large-scale formations can have 'effects' upon the other entities, and that coding all this up will enable deep linkage of mixed-methods data. The particular example I gave in India and Bangladesh showed two findings: first that the invisibility of women's work using official discourse about employment was belied by the high visibility of women's work in semi-structured interviews. Indeed (secondly), women's work was much appreciated and was assigned social value by participants in discourses around work. The methods we used included coding key terms, comparing them and grouping them into key topics, noting competing discourses, and coding up intertextuality. Re-theorising the situation could occur as a synthesising moment.

This chapter thus underpins the use of qualitative data in mixed-methods research. Such data and their transparent codings are of great use for reflexive theorising, and they enable both internal conversations and rehearsing of interpersonal arguments in public settings (Archer 2009, 2014).

Throughout this chapter, retroduction proved to be useful. Induction itself is not the only way to arrive at findings. Deductive hypothesis testing was put into a basket of its own—shown to be useful but not all-encompassing. Deductive hypothesis testing was posed as a method of falsification, which in its origins was aimed at avoiding a fallacy of verification. But to run hypothesis tests may help an audience to see what claims they must rule out. It thus facilitates discussion in the grey zone where arguable, warranted claims compete for attention.

Review Questions

A text is not the same as a corpus: Name two differences.

Explain and give examples of (1) discourse, (2) discourse analysis, and (3) hermeneutics.

When do interviews help give or generate knowledge of social institutions? Can you give three examples of interview questions that might generate new insights?

Is it necessary to examine competing discourses to engage in a theoretical debate about your current research topic? Say why, or why not.

Summarise a current theory debate and how discourse analysis might contribute to developing one of the positions in that debate.

Give details of one coding procedure, and state what two presumptions are commonly made in coding up the meaning of given terms within a qualitative textual corpus.

Further Reading

Options for research design about macro entities are helpfully set out by Blaikie (2000) and Blaikie and Priest (2013), giving examples to illustrate how to avoid atomism and excessive individualism. A sociological account of causality debates is found in Lopez and Scott (2000). Layder (1993, 1998) is helpfully explicit about layers and what is meant by meso and macro phenomena. Lewis (2000) considers causality in detail. Maxwell (2012) offers guidance for qualitative research from a realist perspective.

Appendix

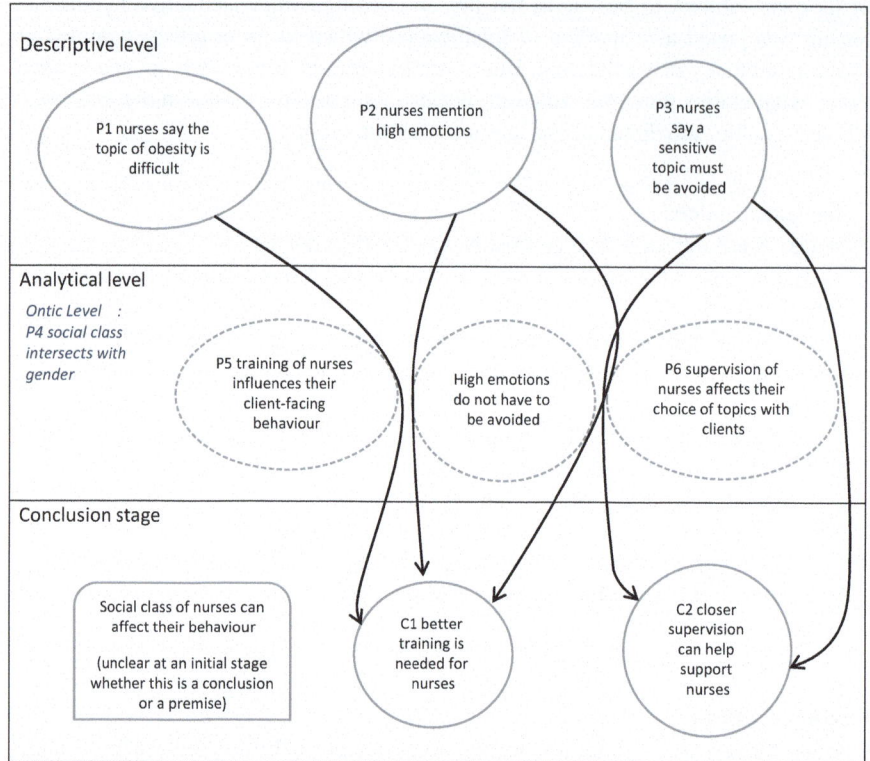

Fig. 8.1 Summary of Arguments in an inductive content analysis paper. (Developed from Mäenpää and Vuori 2021). Notes: Induction is shown by lines drawn from premises P towards conclusions C. In addition, matters for reflection and theoretical elaboration are shown as ontic level, in dotted-line circles, and in the rectangular box. One role played by quantitative evidence might be to clarify aspects of this rectangular box. The conclusion would then develop a clear link with additional evidence

References

Archer, Margaret (2009), *Realist Social Theory: The Morphogenic Approach*, Cambridge UK: Cambridge University Press.

Archer, Margaret (2014) *Structure, Agency and the Internal Conversation*, Cambridge University Press.

Bhaskar, Roy (1997) *A Realist Theory of Science* (2nd ed.). London: Verso.

Blaikie, Norman, and Jan Priest (2013) *Designing Social Research. 3rd Edition*, Cambridge: Polity Press (2nd ed. 2009).

Blaikie, Norman W.H. (2000) *Designing Social Research: The logic of anticipation*. Cambridge, UK; Malden, MA: Polity Press: Blackwell.

Bowell, T., & Kemp, G. (2015) *Critical Thinking: A Concise Guide.* 4th ed., London: The University of Chicago Press.

Carter, Bob, and Alison Sealey (2009), "Reflexivity, Realism, and the Process of Casing", Chapter 3 in Byrne and Ragin, eds., The Sage Handbook of Case-Based Methods, 2009.

Charmaz, Kathy (2014) *Constructing Grounded Theory,* 2nd ed., London: Sage.

Danermark, Berth, Mats Ekstrom, Liselotte Jakobsen, and Jan Ch. Karlsson, (2002; 1st published 1997 in Swedish language) *Explaining Society: Critical Realism in the Social Sciences,* London: Routledge.

Fairclough, Norman (1992) *Discourse and Social Change,* Cambridge: Polity Press.

Fairclough, Norman (2001), *Language and Power,* London: Orient Longman.

Flyvbjerg, Bent (2011) *Making Social Science Matter: Why Social Inquiry Fails and How it Can Succeed Again,* Cambridge, UK: Cambridge University Press.

Gibbs, Graham R. (2002), *Qualitative Data Analysis: Explorations with NVivo,* Buckingham: Open University Press.

Jeffery, Patricia, and Roger Jeffery (1996), *Don't Marry Me To A Plowman! Women's Everyday Lives In Rural North India,* London: Routledge.

Layder, Derek (1993) *New Strategies in Social Research,* Cambridge: Blackwell Publishers. Also Repr. 1995, 1996, Oxford: Polity Press.

Layder, Derek (1998) "The Reality of Social Domains: Implications for Theory and Method", ch. 6 in May and Williams, eds. (1998), pp. 86–102.

Lewins, Ann, and Silver, Christina (2007) *Using Software in Qualitative Research Step-By-Step Guide,* London: Sage Publications.

Lewis, P. A. (2000). "Realism, Causality and the Problem of Social Structure", *Journal for the Theory of Social Behavior* 30: 249–268.

Lopez, Jose J. and Scott, John (2000) *Concepts in the Social Sciences, Social Structure,* Buckingham: Open University Press.

Mäenpää T., & Vuori A. (2021). "Broaching Overweight and Obesity at Maternity and Child Health Clinics". *International Journal of Nursing Practice,* e12958. https://doi.org/10.1111/ijn.12958, accessed May 2021.

Maxwell, Joseph A. (2012) *A Realist Approach for Qualitative Research,* London: Sage.

Neff, D. (2009), PhD, University of Manchester.

Olsen, Wendy (2006), "Globalisation, Liberalisation And A Paradox Of Social Exclusion In Sri Lanka", chapter in A.H. Carling, ed., pages 109–130 in Globalisation and Identity: Development and Integration in a Changing World, London: I.B. Tauris.

Olsen, Wendy (2009a) "Non-Nested and Nested Cases in a Socio-Economic Village Study", chapter in D. Byrne and C. Ragin, eds. (2009), *Handbook of Case-Centred Research Methods,* London: Sage.

Olsen, Wendy (2009b) "Beyond Sociology: Structure, Agency, and Strategy Among Tenants in India", *Asian Journal of Social Science,* 37:3, 366–390.

Olsen, Wendy (2010) "Realist Methodology: A Review", Chapter 1 in Olsen, W.K., ed., *Realist Methodology,* volume 1 of 4-volume set, Benchmarks in Social Research Methods Series. London: Sage. Pages xix–xlvi. URL https://www.escholar.manchester.ac.uk/api/datastream?publicationPid=uk-ac-man-scw:75773&datastreamId=SUPPLEMENTARY-1. PDF, accessed 2021.

Röhwer, Götz (2011), "Qualitative Comparative Analysis: A Discussion of Interpretations", *European Sociological Review* 27:6, 728–740, https://doi.org/10.1093/esr/jcq034.

Sayer, Andrew (2000) *Realism and Social Science.* London: Sage.

Scott, James C. (1977) *The Moral Economy of the Peasant: Rebellion and Subsistence in Southeast Asia,* Ithaca, NY: Yale University Press.

Tashakkori, Abbas, and John W. Creswell (2007), "The New Era of Mixed Methods", *Journal of Mixed Methods Research,* 1: 3, 3–7, https://doi.org/10.1177/2345678906293042.

Teddlie, Charles, and Abbas Tashakkori (2009) *Foundations of Mixed Methods Research,* London: Sage.

Part III

Interpretation and the Validity of Research

Interpretations, Meanings, and Validity in Mixed-Methods Research

9

Multiple strands of argument can offer possible foundations for validity, together creating a grey area where discussions about 'true' claims can take place. This chapter focuses on issues like this, which are known as the 'epistemological' issues that refer to knowledge, validity, and multiple interpretations. This chapter moves from describing broad issues of ontology as it relates to knowledge to assigning a specific role to values in setting up the validity of an argument. Finally, offering a contrast of a falsification with a warranted argument. These elements of epistemology can be combined with each other, because of society's fundamentally 'open' (i.e. changing) social systems. Much more can be said about epistemology (see Further Reading). There might be false claims, about which there is evidence that we can use to point to alternative arguments. There might be claims in the grey zone which are debatable or comprise part of an ongoing contestation; and there might be claims which are true. By allowing for a grey zone, this chapter urges that a simple 'fact vs fake news' distinction may not be viable.

9.1 Truth Is Not Simple in a Complex Society

▶ **Tip** This book has not argued for a relativist concept of truth in which any statement can be true if, or because, its stakeholder or speaker thinks it is true. Instead, the statement rests within discourses which are operating to help agents communicate in a complex society. Respecting what people think implies learning what the meanings are of statements made in existing discourses. A challenge to a particular statement, however, can draw upon points made in a separate discourse (perhaps one from a scientific source, or a social theory or a social-policy perspective).

Rather than being able to refer to a realm of 'truth' via a separation of objective facts from subjective knowledge, a starting point is that the objective and subjective

W. Olsen, *Systematic Mixed-Methods Research for Social Scientists*,
https://doi.org/10.1007/978-3-030-93148-3_9

aspects of human knowledge overlap. One can distinguish information from knowledge. Information refers to symbols which represent or summarise a situation, which are disembodied from human discursive acts; whilst knowledge is the embedded and accessible core of practical abilities in humans and other agents. By implication, information helps discursive acts to operate since information is embodied in human society; it has been created within human society. 'Discursive' acts mean those communicative acts having some intention or meaning in a society (see also Chap. 8). Using this wording, subjective meanings are associated with human and institutional agents, and information is created by these agents, such as governments. Making a distinction between knowledge and information is one step towards clarifying that knowledge has both subjective and objective aspects simultaneously. Some objective aspects of knowledge were discussed in other works on philosophy of science, where moral realism, realism of entities, and discourse itself were examined closely (Outhwaite 1987; Potter 2000; Porpora 2015). For most researchers, it is acceptable to realise that a subjective aspect exists for many social situations; but the evidence-based knowledge movement has created pressure to give evidence to support any knowledge claim that we want others to accept. The evidence might include a range of types of evidence and argumentation. This chapter will consider aspects of that argumentation.

By allowing the overlap of subjective and objective aspects of knowledge, one may become more reluctant to discuss ultimate truth, because most assertions can be seen as subjectively influenced. One may also become more interested in meanings and intended meanings. In spite of intentions, meanings of an utterance or a social theory can get out of hand, straying beyond what is intended. (Backlash against feminist statements might be on example; similarly there can be unintended backlash against the laws that promote social equality.) A grey area opens up where one person's truth is someone else's fake news. Another way to say this is that scientists creating a 'school of thought' may establish as facts a series of grounding points, such as a theoretical perspective, measurement traditions, and the geographical and temporal scope of claims, within which it is possible to believe that foundations of 'truth' exist. As time passes, or when new stakeholders arrive at a discussion, the grey edges of this zone of normal truths become larger and are recognised as giving scope for debate and disagreement. Mixed-methods research helps to pursue basic democratic principles of listening to competing arguments, and democratic approaches in daily life will tend to imply a need for pluralism in research. Here, the word pluralism has two usages. First it can be used to refer to allowing for multiple theories. One example is that workers labouring in rural India or Pakistan could be examined using a mix of human capital theory, peasant labour supply theory, demographic theory, and power theories around landlord/tenant relations (Olsen 2006). The pluralist has a larger canvas to work upon than someone embedded in a single school-of-thought conversation.

Secondly, another meaning of 'pluralism' has arisen in the context of the mixed-methods debate: methodological pluralism, which means mixing methods or putting together diverse methods into one study. This kind of pluralism has had some good coverage in the methods literature (see Chaps. 1 and 2; Downward et al. 2002).

For those who want to amend a paradigm, move forward in a hot debate, or solve an apparent paradox, this kind of pluralism can generate innovations. Methodological pluralism has been promoted throughout this book. See also Further Reading for this chapter.

The act of 'falsification' (which is explained more later in this chapter), which is taught in many textbooks on social statistics, may morph towards acts of 'challenging' and justifying claims. Any message can carry multiple meanings, and most phrases or sentences will tend to be open to a range of interpretations. For example, a hypothesis that A is correlated with B will carry nuances of meaning to those who read A and B conceptually as elements of wider discourses. Within a particular discourse, A and B are concepts which have normative overtones and which carry their historical origin's stereotypes towards the future. A simple example from Chap. 4 is that aid was construed as funding for social assets, and the quantum of new community assets, such as public toilets or water supply, was correlated with improvements in household-level poverty levels in Egypt (Abou-Ali et al. 2010) The concepts 'aid' and 'poverty', when unravelled in a multidisciplinary development-studies context, refer to much more than quantities of money flowing into and out of projects and households. They also tap into power structures, systems of control, international post-colonial influence systems, and other open systems composed of meso and macro elements. So it is very hard to simply test one hypothesis from amongst this complex nexus of meanings and assertions.

To take a different example, in economics, the role of non-pluralism meant for some decades that an economist building a 'utility maximisation model' would be incidentally promoting a particular atomistic, heteronormative, commercialistic, and/or individualistic world-view. Heterodox economists are now trying to create alternatives to this simplified view of economic actors (Lawson 2003). The very idea of a simple correlation suggests atomistic units—firms, farms, or people—whose features are fundamentally comparable. Simple and multiple regression model estimates are fundamentally based upon correlation and therefore have this ontological tendency. Experts acknowledge that meso entities like production standards and regulatory regimes, and macro entities like the property rules and legal system, do matter. These meso and macro units do also influence the trajectory of A and of B and can create changes, over time, which make arguments about the correlation 'confounded'. I have argued that a wider purchase on the situation can be obtained by studying micro, meso, and macro units, implying a need for mixed methods to handle the aspects that survey data or quantification may not cover well. More than just supplementing quantitative research, we need an integration of theory that allows for micro, meso, and macro objects—all of which are changing over time. In summary, neoclassical economic theory is one set of hypotheses which need not just falsification, but wholesale revision and replacement with improved theories of how economies operate.

It was made clear in the regression Chaps. 3 and 4, and in the factor analysis Chap. 5, that causality across the levels was important. A new action by a single micro agent could influence a macro object (as seen in the study of a 'results-based' reward system, or Results-Only Workplace Environments (ROWE) by Kelly et al.

2011). Furthermore, a macro situation or macro entity can influence some micro individuals more than others, as presented in a structural equation model with an asset index or a Gini coefficient among the independent variables. The asset index or Gini coefficient can represent inequality. (Asset indices are measured at the level of each unit whilst the Gini is an aggregate measure for groups of cases, such as a city.) Interaction effects can then cut across the levels by combining the effect of a unit-level variable upon the outcome, moderated by a city-level variable such as inequality. These cross-level interactions will be challenging to represent in a statistical model, although we can try. Fundamentally they reflect the open nature of the human society and the ability of different parts of it to affect the future trajectories of it all.

In this context, Chaps. 1 and 2 urged a form of strategic structuralism. There are many meanings of 'structure', as discussed there. Strategic structuralist approaches differ from those relying upon a more deterministic form of structuralism. Deterministic structuralism might say that a social structure definitely and necessarily has a certain impact upon the participating agents, for example the Care Home management structure affects all the patients. The weaker forms of structuralism would see the Care Home management structure as influenced by a range of historical and agency factors, and it would see patients as diverse in their capacities, too. The 'strategic structuralist' would consider all these possibilities, and then gather data with a view to improving some outcomes in the long-term Care-Home context. Taking factor analysis as another example, I have questioned whether the use of factor analysis implies any deterministic approach to causality. For instance, if we make a simple forecast that one ethnic group might have lower average values on a well-theorised factor for a social norm, the hypothesis is likely to prove unwarranted. Many contextual conditions affect the range of answers that members of that ethnic group and other ethnic groups give to the manifest variable questions. In brief, there is diversity in the scale index within ethnic groups as well as between ethnic groups. The social structure does not simply determine social norms or the values people hold. In recent years it is considered normal to keep intersectionality of diverse structures in mind, and intersectionality also implies low levels of deterministic causality (note that falsificationists are rarely working on intersectional inequalities, but there is some overlap). The structures are usually considered to range from age, sex, and ethnic group to social class, religious group, and country or language of origin.

In using the word 'structures' I am instead arguing for a weak form of structuralism, which allows for many intervening and cross-cutting factors. Structural factors might generate mechanisms that are influential, but not deterministically causal on final outcomes. Mechanisms-based reasoning was surveyed earlier and it includes the Context-Mechanism-Outcome approach (Pawson and Tilley 1997; Pawson 2013), which is meant to include the possibilities of factors operating at micro, meso, and macro levels at all three stages C, M, and O. It is not meant to imply a super-simplified approach to single causal processes, because in the CMO literature it is stressed that complexity exists and things overlap.

In this book, I pointed out that structured objects have emergent properties, which cannot be reduced to their component parts. We have to make some choices to develop a simplified argument about the complex social structures and agents; therefore, research-based arguments may be developed as part of the authors' voice, and become a part of strategies for social change. Thus whereas the people who hold that there is a 'schism' of qualitative versus quantitative research traditions see qualitative methods as focused on voice and authorised to discuss agency, I have argued that the 'schism' is only artificial and weak, and that with integrated mixed methods all researchers are in a position to discuss voice and agency. Like most qualitative researchers, I also argued that the researcher themselves have an ethics of self-reflexivity and consciously try to do good, or at least not do harm.

One epistemological implication of the above argument is that pure description is possible, but that the elements and concepts used in any particular description may be open to challenge. If a doctor writes a summary of a patient's condition, they are writing what they consider valid but that is not necessarily the same as writing true claims down. The truth of the claims would have to be checked, and ideally subjected to falsification and/or retroduction. The scepticism that results is wide-ranging. For example, then the methods used to discern structural features should be sensitive to the dynamics of systems and the intentions of the agents in each system. In studying the water industry, one might take into account the professionalisation and/or corrupt practices of water engineers in a given region. In this chapter I take up what happens to 'facts' when there are multiple stakeholders in these dynamic systems.

I have referred earlier to *mezzo-rules* (part-rules). Morgan and I (2010) developed the idea of mezzo-rules to help us talk about rules as reference points without committing to a reifying concept of hegemonic rules. Definitions of words, like 'structure', 'norms', 'shaming', and others, are themselves rules which we should also see as mezzo-rules. The definitions are open to debate. Constructivists specialise in the nuanced interpretation of multiple meanings of words and other discursive acts. Researchers can avoid assuming deterministic rule-following in general, as well as in relation to the terms they use. Thus a 'marriage' was broached in Chap. 1 as having many meanings, many rituals attached to it, and therefore being open to redefinition. It is possible to study rules and norms using discourse analysis, or other methods, alone and in combination with quantified evidence. This book suggested that the purely qualitative mixed methods might miss some opportunities for intriguing descriptions, and one example is provided below.

Suppose the focus-group researchers (Mäenpää and Vuori 2021), mentioned in Chap. 8, looked into the social structural patterns of obesity in Finland. Using a national survey, they can show the social-class differences in average body mass index of adult women, and in children. The point was made informally that overweight was more common among the lower socio-economic groups. The risk of a woman of 'lower' socio-economic status arriving overweight as a client in a pregnancy clinic is much higher than for women of middle to lower wealth households in Finland. The issue of broaching a sensitive topic like diet and exercise with clients might be coloured by the social status and economic power of clients, and nurses, in their socio-economic group of origin. Therefore, the social class of nurses will matter. As broad contextual matters, the country's economic inequality and its policies on equality will then be important for this research. The findings can be enriched by developing the social-class angle when interpreting the focus-group data.

Researchers thus might link qualitative analysis with quantitative survey-based tables. Researchers can link 'critical discourse analysis' with action research, or grounded theory with keyness analysis, to gain a sense of the dynamism that exists when a society has mixed, non-hegemonic, diverse social rules. In summary, because society is open and changing, and people follow a range of mezzo-rules, the researchers have to develop an awareness of how things are changing in the surrounding contextual society for any given project. In this context it is hard to argue that any particular claim is an 'objective fact'. The phrase 'objective facts' gives a flavour of permanence whereas human society is usually changing and changeable. It is better to argue that a claim is a warranted claim.

The UK government has given advice for policy evaluation, suggesting that a variety of methods can equally be used to achieve a valid set of claims (Majenta Book supplement, Quality of Qualitative Research, see URL https://www.gov.uk/government/publications/government-social-research-framework-for-assessing-research-evidence, accessed 2019). The UK government's underlying idea is realist in that diverse methods will tap into the same real processes (Dow 2004). An example of linkages at multiple levels is Cross and Cheyne (2018). Here is how they described the linkage-making process during their research planning.

> Complex interventions … require a methodology that captures complex dynamics. Consequently a realist-evaluation-informed case-study approach was adopted across three contrasting health boards, comprised of: (1) interviews with women … with heightened risk, (2) a sample of maternity care professionals responsible for implementing the policy, and (3) document analysis of policy guidance and training materials. (Cross and Cheyne 2018: 425)

They also developed deep, substantive linkages across data sources when drawing their conclusions.

Making these linkages is going to work as long as your project is managed well, has some time boundaries, and does not exceed its own cost limits or time limits.

The methods in this book have departed from the set of exploratory data science methods known as 'statistical learning' and from standard hypothesis testing methods. Instead the book built upon realist and retroductive foundations (less attention is given here to constructivist approaches). These methods are a little different from 'text mining' or data mining, as described in textbooks on data science (Silge and Robinson 2017). In particular, the partly computerised coding of 'discourse elements' can be used by a variety of schools of thought: critical discourse analysis, political economy, business and management, education, social constructionists, and many other researchers (e.g. NVIVO software; see Gibbs 2002; Lewins and Silver 2007). Using the alternative of 'statistical learning' alone, that is pattern discernment by a computer algorithm, it is unlikely that the holistic conceptual framework we need to arrive at would be created. Such frameworks require human input not just artificial intelligence.

Table 9.1 suggests how realism and reality are embedded into the interpretation of qualitative data. This table draws upon the concept of situated knowledge (Smith 1998), where the interpretation of qualitative materials can be firmly rooted in warranted arguments and is grounded in a certain place and time, subject to critiques that would inevitably arise from stakeholders in other places/times or who rest upon other theoretical frameworks. Table 9.1 does not rest upon strong social constructivist arguments but is consistent with much of interpretive social science (see e.g. Kvale 1996).

Table 9.1 shows that the empirical domain, where our interview data or documents and visual evidence sit, does not exhaust the 'real'. Things can be real and have effects, or can be intertwined with other social entities, regardless of whether the data shows them to be so. One implication is that research results are fallible. Fallibility arises when we realise the 'real' is so complex that our statements about it, such as describing an actual event, are likely to fall short of being an exhaustive or even accurate truth. They are representations. As such, some scholars go to the extreme, of thinking that the social phenomena which we can know about are merely social representations and have no intransitive reality. There is no need to go so far.

A transitive aspect of reality is the part or whole which is changed by our exploration of it, for example a conversation. An intransitive aspect of reality is the part which is pre-existing, such as older memories, habitual reactions, or systemic properties of a pre-existing structure. See also Archer et al. (1998) for details of transitivity versus intransitivity in competing approaches to knowledge.

One key realist argument is that there are some pre-existing entities, and therefore some parts, aspects, dimensions, and historical facets which are non-transitive to our knowing gaze. The role of royalty in medieval economic power systems, rooted in lineage norms, ethnic and class structures, and taxation norms, would be an example of something that may have pre-existed the current power system. (An

Table 9.1 Values and validity in two forms of social science

Traditional science approach	Situated-knowledge approach
Validity	Makes reference to recorded evidence
	Critical assessment of evidence
Replicability	Makes reference to contexts
	Makes reference to reality not merely to theory
Reliability	Offers transparency by offering some evidence for independent scrutiny
	Sophisticated and/or systematic data recording
Social science mimics natural science (naturalism)	Appreciates diverse standpoints as special feature of social science (reflexivity) and of society; the social scientist has a specially wide and deep knowledge
	Plurality of theories, and critical approach to theories of change
	Depth ontology involving nested and non-nested cases
	Authentic voices

For details see Olsen (2012). For a discussion of situated knowledge see Smith, ed. (1998)

alternative view is that the current system is rooted in the medieval system and that their evolution makes them two parts of the same changing system.) The idea of 'intransitivity' is that even while we seek knowledge about the thing, it does not change, and it is our knowledge which is changing over time. Realists mainly point to events and the 'actual' domain for examples of intransitive entities. By contrast structures are nearly always undergoing change processes, and are likely to be transitive. Entities can be partly intransitive and partly transitive. The historical textbooks on colonialism and on slavery illustrate this. Looking at multiple books, which offer a changing coverage of actual situations, it is arguable that the colonising actions and slavery events of the sixteenth to nineteenth centuries can no longer be changed by our descriptions. As such, the intransitive mechanisms, events, and experiences might be hard to tap into. Therefore, we use evidence as traces of the real. At the same time, recourse to evidence does not resolve the deep question of what discourses we should use to analyse and unpick the colonial experience. In Chap. 8, this book illustrated some of the ways transitive knowledge can be gained through qualitative analysis. I gave examples of some real mechanisms that one can discover more about using discourse analysis. These correspond to rows 1, 2, and 3 of Table 9.1.

The meaning of the key terms here is set out in other works on realism (Archer et al. 1998). Mechanisms would be the capacities of things that make it possible and also likely that they will generate an effect or a change in other things. Meso-level mechanisms are aspects of things at a cross-cutting or encompassing level, such as regions in the UK, the national head offices of a firm (contrasted with its physical units in 25 places), and the courts (in which court cases are nested). Meso-level entities have features not held by the units that are nested within them. Finally, macro entities are wholes not divisible into their parts in an easy or comprehensive way: the 'government', the 'state', the 'corporate brand', and the 'legal system' would be macro entities.

9.2   Epistemology for Late-Modern Mixed Methods

There could be several competing 'valid' arguments about one mechanism. These arguments would co-exist in a democratic or pluralistic space of dialogue, and may not all be equally good. The arguments might be spread out on a spectrum: feminist to anti-feminist, or right-to-left wing, or market-to-government oriented. They will not be evenly spaced, some arguments being far apart from a core of consensus knowledge, and others being variants on a common theme. One overall 'theory' comprises a few different claims, and each claim in turn is based on its premises and its data and reasoning. For individual people, holding to a theory may also rest upon visceral feelings, experiences, loyalties, emotions, beliefs, or memberships. Since arguments are complex, arguments for and against one theory are likely to co-exist while the debate goes on about which is better/worse and what purposes they serve. In this sense, theories are in a grey area. They are epistemologically contestable.

Table 9.1 provides a list of the dimensions of a good knowledge claim, and hence of highly valued knowledge. These include giving evidence, providing some context, transparency of data and methods, a sophisticated approach to truth, and pluralism. Yet in spite of wanting to promote tolerance of multiple views, it is still likely that theories and explanations need a good foundation of some kind. The foundations can be asserted using either logic or metaphysics. For instance in realism one of the foundations is the pre-existing nature of some of the world, prior to our research about the world. A phenomenologist might, instead, offer a metaphysical claim or a constructivist assertion to get started with foundations for valid findings. The five key tenets of valued knowledge, shown in Table 9.1, apply to all kinds of research and to a variety of types of claims. Three levels of claims are mentioned in the table: concrete, intermediate, and more abstract claims. The middle column presents a scientific naturalist approach at an abstract level, while the last column presents a more situated social approach to knowledge and uses concrete examples (Smith 1998). Situated knowledge works better for doing discourse analysis. Naturalism by contrast would be suggesting there is 'God's Eye View' upon a discourse, whereas in reality we analyse one discourse from within our situated locations in other discourses.

I have not dealt with the interesting history of terms like naturalism or structuralism here. See entries in the *Stanford Dictionary of Philosophy* (https://plato.stanford.edu/entries/hume/ or https://plato.stanford.edu/entries/structural-realism/ for example, accessed 2019). Such terms as 'realism' have held varying meanings over a period of centuries.

There could be two or three levels of claim. To illustrate: the phrase 'the capacity to pollute' represents the capability that a factory or nuclear plant has to release damaging pollutants. Discussing the mechanisms of releasing pollutants, we get in a debate about the power to release and the power to inhibit the releases. These claims about power or capacity to act fit in a particular discourse. At the highest level, it may be that the power of neoliberals is underpinned by the repeating of their dominant discourse tropes about market externalities such as pollution. My example here is rooted in the UK and Indian situations, but this dominant discourse has been exported and has influenced many other debates about a range of pollutants. So far, there is little successful impact of marginalised discourses about the claim 'pollution is intrinsic to private commerce'. This example is rooted in a study of political economy, for which some methods of research can be found in Touri and Koteyko (2014). Thus the idea that commercial firms must change their orientation to externalities, and not just filter, export, or bury pollutants, is a marginalised discourse. It argues that a tendency to pollute is at present intrinsic to private commerce in the UK and India. A critique of private commerce and particularly of unregulated manufacturing and power plants arises from within the marginalised discourse. A social scientist must be aware of many pros and cons when deciding what tactics to use in constructing arguments about such a situation.

My example illustrates that a scientist can argue in favour of profound changes and not be restricted to arguments about incremental changes within existing discourse structures. In the pollution example, a researcher need not just choose what per cent of externalities the private firms should bear. They can argue for a complete revision of theory, or for a re-structuring, or for a re-fashioning and a new vision of the polluting industries. Alternatively they may argue about regulation and focus less on the pollution process and its costs, but more on the structure of regulation. Touri and Koteyko (2014) argue for a close link from evidence to findings in such research projects that challenge dominant neoliberal discourses. They advocate induction and description. My case goes further because a scientist is motivated to achieve outcomes strategically and may decide to be creative or innovative, not just descriptive. To illustrate, Morgan analyses the financial crises of 2008–12 using forward-looking contrastive explanations (Morgan 2013).

Taking the non-naturalistic stance, we gain traction for new knowledge claims but we then compete inside the grey area where the standards of knowledge are not simply 'validity', 'replicability', and 'reliability'. Table 9.1 clarifies further the activities of a situated-knowledge expert, who is a scientist but not naturalistic.

These tenets offer a strong contrast to the ideas of Popper (1963) about his proposed methods of validity testing.

To summarise some epistemological guidance arising from the realist approach, five of the key epistemological aspects are the strategic use of evidence; presenting context; transparency; a sophisticated approach to truth; and pluralism. Mixed methods are a good approach to research design based on these tenets (see Table 9.2).

Table 9.2 Epistemology in critical realist social science

Valued tenets of realist approach	Abstract examples	Concrete examples
Evidence	Makes reference to recorded evidence Critically assesses evidence	Shows a QCA data table Explicitly mentions all stages of sample selection Provides transcripts Discusses weaknesses of interview technique
Context	Makes reference to contexts Makes reference to reality not merely to theory Goes beyond the description of events	State what the geographic reference point is State what social groups are covered, and the time period Note topics that were not given attention, e.g. Jaffna Peninsula in Sri Lanka was outside scope
Transparency	Offers some evidence for independent scrutiny Sophisticated and/or systematic data recording	Transcripts are put in a data archive Sample quotations are given in context Case-study summaries are put in a repository The quoted cases are numbered and linked to scattergram markers

Table 9.2 (continued)

Valued tenets of realist approach	Abstract examples	Concrete examples
Sophisticated approach to truth	Appreciates diverse standpoints as a special feature of social science (reflexivity), and as a feature of society Sees the social scientist having a wide and deep knowledge Assumes a multi-voiced society, which creates grey zones of potentially true statements	Presents ideal types with quotations and local detail Shows how banks and customers differ in their narratives about finance Includes stakeholder analysis Does action research on empowerment through access to banks as alternative credit sources
Pluralism	Shows plurality of theories Takes a critical approach to theories of change	Different socio-economic contextual explanations are contrasted, e.g. with demand/supply theories
	Applies a depth ontology involving nested and non-nested cases Exhibits authentic voices	Uses multilevel models alongside truth table and/or discourse analysis Gives direct, verbatim quotes (not third-party summaries)

The whole framework of epistemic values I have set up here is not consistent with parts of the tradition known as falsification. I will explain this tradition in some detail then offer an alternative approach. Both are useful, but they do contrast strongly. It may be difficult to use both in the same project, but there may be good reasons to try to combine them so that results suit a wider range of audiences. I would also argue that the hypothesis-testing approach is too narrow and less creative, and indeed hides what is creative in a research design. Therefore I offer in Sect. 9.3 the falsification method and in Sect. 9.4 the alternative method which most of this book has presented: an approach where retroduction is admitted to be a useful way to look deeply into the explanations of outcomes.

9.3 Falsifying Hypotheses: Possible and Desirable, but Not Necessary

Popper argued no claim in science is valid in itself. He argued against trying to prove your own beliefs. He suggested that a theory and its hypotheses can be shown to be **not-disproved** and that this would be a valuable step in logic. Below, I summarise this logic, and Box A below can be contrasted with a competing logic found in Box B later on.

A Falsification Logic: Optional and Scientific (Box A)
 – Reading the literature, we address a theme within that literature.
 – We gather evidence about it using existing phrases and measures, so that evidence becomes objectified and transcribed, and can be assessed rigorously and by multiple researchers.
 – Our aim will be to generate reliable claims.
 – If data X and data Y are found, comprising surveys, interviews, texts, sound recordings, and other evidence.
 – And we find relation X-Y in these data.
 – We can discern claims about causality in the literature H1 and H2.
 – Then we can assert correctly that $X \rightarrow Y$ or $Y \rightarrow X$.
 – Yet we can never assert which one of these is true.
 – We can make other claims such as $P \rightarrow Y$ and $X \rightarrow P$ and so on, but not being sure which is causal, we seek data to show which causal pathways are *not* the case.
 – Lastly, omitting all falsified claims, we are left with the one (or a few) remaining scientific claims.
 – Untestable claims do not belong in the basket of non-falsified claims as they lie outside science ('The Demarcation Criterion').

Note: P=Premise; X, Y = variates or text chunks; H=hypotheses.

The supporters of Popper since then have advocated testing and re-testing; replication of experiments; and reliability as criteria for external validity. Their approach often assumes a uniform world with homogenous agents in it. The approach is fundamentally atomistic. It assumes away regional cultures and local social variation in culture. It could be said to be lawlike, but laws of society do not exist, so in this sense it is founded on a poor quality ontology. The term nomothetic is used to describe the classic orientation towards laws, which is a specific general ontological assumption. This is why the Popper approach is very closely allied to the nomothetic hypothesis-testing tradition.

By assuming laws exist, the task could then be to discover them. If we assumed they were not laws, but instead *mezzo-rules* and social norms, then we might do the research differently. Because my research experience has tended to show me many examples where social norms are broken or confused, and a plurality of norms compete, I tend to think of society as not lawlike. In such a case, testing hypothesis based on lawlike assumptions creates an incoherent logic. It may be better to assert that socially normal patterns are reproduced, but also influenced towards change through a variety of actions and that the agents' intentions are important for the actions chosen. This makes a more fluid system the subject of research. It also implies more need for retroduction, and less utility of hypothesis-testing.

To clarify and illustrate the term 'nomothetic hypothesis testing', an example from the SARS-COV-2 virus and its related disease, COVID-19, can be given.

International bodies offer definitions of the virus type, with its acronym, and the disease type with its associated name. Thus, the world may consistently use a word like COVID-19 for a period of some years. The nomothetic approach could argue that the experiments with vaccines intend to create a medical injection solution to reduce COVID-19 risk across the world. The hypothesis that a particular vaccine will work to reduce the impacts of the SARS-COV-2 virus is then subjected to tests in different parts of the world. Many medical and statistical voices rehearse the arguments of hypothesis testing, quoting numerical evidence, to state which vaccines work and how well. The 'replication' is re-testing a vaccine in a new environment to see if it works as well there as in the first experiment. The 'reliability' test is that a test result should be as robust and significant in one environment as in another, or over time. The 'nomothetic' part is the assumption that a dose-effect relationship exists worldwide. Would that relationship be diverse in different parts of the world, or would bodies be differentiated in their response to the vaccine? Is the virus having different effects in different places, and why? These two questions are less likely to be asked in the experimental vaccine tests because of their nomothetic hypothesis-testing basis. That research tends not to be done in either a retroductive or inductive way. People's beliefs about how vaccines may have side-effects are not considered at all relevant to the effectiveness of the vaccine. The term 'effectiveness' is defined etically (from outside the culture), and medical discourse routinely ignores all beliefs of all experimental participants even though those beliefs are diverse. They are considered irrelevant. Thus, the nomothetic tradition is lawlike and atomistic, and is used in a research area like human biology where exceptional human bodies are considered outside the remit of the main research campaign. Bodies with tuberculosis, those who are children and those who have already had the COVID-19 disease may all be left out. The nomothetic tradition has norms in it and tends to reproduce those norms, and one makes narrow tests on a given theoretical platform that is not, itself, tested.

However the impact of Popper during the period after World War II was immense in Western countries' science establishments. Falsification was an achievement at the time. The method of nomothetic testing was, in Popper's writings around 1955–65, as far as a scientist can get. Science, for him, was demarcated by the attempts at falsification. This was known as the demarcation debate: what bounds science away from non-science? In this the scientist was meant to be very open-minded. They can consider alternatives Not-X, such as, new Xs, such as 'the SARS-COV-2 virus causes new syndromes which need to be named as sub-types of COVID-19', and whether the model built up by others, and not falsified, could perform better or worse than a new model, also not falsified. In the case of COVID-19, it might be asserted that we need to test whether a rapid obesity-reduction programme could reduce the SARS-COV-2 virus negative impacts as much or more in the same time period as the vaccines. Because obesity is associated with heart and lung difficulties, and reducing obesity could be associated with improved breathing, such a programme of research might find it not falsified that 'reducing obesity reduces negative effects of this virus'. In summary, one can test narrow hypotheses within a given scientific theory, or one can test whole competing theories, using data that fits the particular project one is setting up.

What was considered non-scientific was to engage in a battle of beliefs, such as a discussion of beliefs about the vaccines or about the ethics of vaccine side-effects. Any scientific discussion would have to be based on evidence, and the use of evidence would have to be guided by competing, explicit hypotheses.

In Popper's time, instead of being considered as a metaphysical question, the demarcation problem was to be solved by giving an explicit methodological answer. The way we set out a 'research design' as an explicit advance plan is based on this premise that it makes the research more scientific, and hence more valuable or more valid. In more recent decades it could be said that some forms of economics, too, have a clear boundary to what is acceptable research, and that this boundary is policed via methodological rules. However these boundaries are contested (Downward and Mearman 2007). An interesting use of mixed methods in the economics of work was presented in each of these studies: Ehlers and Main (1998); Hill (2001); and Van Staveren (2014). In each case, the supposedly scientific core theories were questioned and replaced, with alternatives being examined based on alternative fundamental assertions rather than merely by evidence-based hypothesis testing.

This implies that some alternatives to falsification methods are more creative, are perhaps more retroductive, and certainly tend to be very complex.

The falsification tradition is strongly presented in textbooks. Here, validity criteria including tests of goodness of fit are often presented from a deductivist point of view. They seem to say that 'If Data X, and data Y, and relation X-Y, and claims about causality, then we can assert correctly that either X causes Y or that Y causes X but we can never assert which one of these is true. Omitting all falsified claims, we are left with the one remaining scientific claim'. The good thing about this logic is that it is explicit about where evidence sits in the trail of reasoning. Another advantage is that where X definitely and wholly precedes Y, Y cannot cause X. By a process of stipulating dates and eliminating options, an explanation of Y may be reached.

Yet this logic also has weaknesses. As expressed above, it is heavily dependent on the data, it requires that there is no measurement error, and it would probably require random sampling or good administrative data. The results are not at all dependent on the person interpreting the data. There is no scope in 'falsification' for individual differences in the conclusions. There is no space for discussing competing discourses, or for our carefully considered choice of a discourse terrain in which there are multiple discourses. Popperian reasoning contrasts with Fairclough (1992) and with Chouliaraki and Fairclough (1999). The mildest forms of interpretivism are seen in Kvale's book *InterViews* (1996), where multiple voices can be heard and each has its roots in the local social meanings, social norms, and social acts of communication. These views would not fit well with falsification. It is this contradiction that has led to the idea of a 'schism' of qualitative and quantitative methods.

I can think of three problems with falsification from this interpretivist point of view (all these problems being consistent with realism too):

- The standpoint of the speaker is historically distinct from that of the listener, so a reasoning speaker will carry out research keeping both standpoints in mind. They adapt their interpretation first to their own standpoint, but also then to that of the Other, and then tweak their own standpoint in a reflexive circle.
- The hermeneutic circle applies within the falsification reasoning, so that the meaning of a measure to one person may not match the socially 'normal' meaning, and hence either a clarification must be made, or falsification given up, because there is scope for multiple interpretations of the things recorded in the evidence. (Here by 'measures' I do not mean qualitative measures, but any records of evidence, e.g. texts, documents, or sound recordings.)
- The requirement for agreement among all reasonable interpreters, commonly agreed in falsification, requires agreement on basic discursive norms. But these norms themselves are likely to be hotly contested, for example do we call patients 'users', 'customers', 'clients', or 'people' in a health setting? Each initial step of the falsificationist's work is terribly contested.

Another serious failing of the validity criteria found in Box A is that an entity's features cannot be isolated from its context. It is intrinsically located at a space-time juncture. An example is that the consistent body responses to medication are utterly changed if the person treated has been starving or is extremely thirsty. Their body deviates from the atomistic norm, and the theory of treatments or dose-response estimates breaks down. Starvation or thirst is a contextual condition. The trick of stipulating the dates of measurements and some other statistical adjustments can bely or mask the key contextual factors in a study.

9.4 A Retroductive Approach

An alternative argument could look like the one in Box B, which offers a realist argument including an X-Y association.

A Realist Claim Rooted in a Carefully Constructed Argument (Box B)
- If the real world exists, and if data X and Y are observed, and their relation appears to be X-Y, then we can assert tentatively that either $X \to Y$ or $Y \to X$, or both. The relation can be existential (ontic), causal, or associational.
- It is useful to find out whether any pathways of cause are dis-proved, so we carry out a falsification phase, see Box 9.1.
- We can explore the relation and the underlying situation further. Retroduction is used here.
- We can say a lot more than we would say using pure induction. We can diverge from the wording of the raw evidence.

(continued)

(continued)

- For example we can also examine competing discourses, compare lay explanations with expert theories, and describe how different spheres and domains of life are intersecting. (Example: the articulation of property laws and norms with capitalist system dynamics.)
- We can also derive conclusions about several concrete areas and then start to work on the wide-scope conclusions. (Example: gender studies in labour and banking can lead towards wide-scope analyses of patriarchy.)
- Instead of 'generalising to the population' of atomistic units, we are integrating diverse entities and experiences at micro, meso, and macro levels.
- We do not confuse inference debates about sample-to-population for one locality with the issue the realists raise about the integration of knowledge over space/time.

I stress that X and Y in the above are not just variables or quantified amounts, but can also be statements or information about any phenomena. X could be the photos of many, many social media postings, and Y could be the commentaries of selected groups of viewers upon seeing these postings. X and Y might also be words in a body of qualitative data, X being raw data and Y being coding annotations. There is no need for causality to be sought. I will stress explanation and causality because it has been controversial and it is interesting. But one can also simply talk about associations, describe patterns and types, and conclude about the articulation of systems or sets of things.

I am concerned about the deductivism in the first argument (Box A). In Box B, I offered an alternative. In Box B, I have explicitly offered ways to avoid excessive atomism. There are multiple logics for deriving claims in the realist approach. In so far as a scientist gets accustomed to using the first 'falsificationist' argument form, they may begin to hesitate to comment on anything outside their immediate experience. The problem is that there is great need to comment on wide social trends, political developments, and other holistic aspects in society. To meet this need, we need the expanded logic of Box B. I find it credible to use mixed logics. The scientific-ness of the work relies on sophistication and transparency in tracing how we carried out the research and derived the findings. Thus, instead of the standard 'demarcation of science' criterion used in Box A, we apply a 'warranted argument' criterion. Credibility is a feature of a specific set of claims bolstered by the related evidence and argument.

Many possible arguments could all be warranted, and these arguments may clash with each other. We need more warranted arguments to have stronger, more robust, and useful disagreements. Fisher (2004) introduced warranted arguments as a technical term, and it is now adopted by many writers on critical thinking (see also Fisher 2001; Mearman 2006; Olsen and Morgan 2005). I found it a convincing way to go about setting up a research-based argument.

9.5   Conclusion

In this chapter, I first distinguished what 'epistemology' meant in natural science and in social science. Terms such as transitivity and objectivity were expanded to allow for the partly intransitive nature of some social objects, such as events, and the overlap of subjectivity with objective when humans are gaining or expressing knowledge. An enriched approach to validity then begins to inevitably use pluralism, and I explained two main aspects of pluralism. One is the plurality of theories and another is methodological pluralism. This latter term often refers to mixing quantitative and qualitative methods, or combining multiple methods in general.

In the next section, epistemological tenets were listed and discussed as possible principles. Because there are several of these, the concepts of 'fact' and 'truth' have blended towards a grey area of competing claims. In natural sciences, the focus may be on validity and useful theories, compared with social science where a larger number of tenets of valid and valued knowledge can be brought to bear together. I called the mixing of all these various tenets of valued knowledge the 'situated knowledge' approach, following Smith, ed. (1998).

Finally this chapter has argued that hypothesis-testing can lie within a wider discourse about knowledge, and thus it is itself contested. Two typical argument types were set out. One is a realist approach to making claims and examining evidence with due regard to falsification. The other is a Popperian falsification approach which posits deductive validity to the claim that a point that has not been falsified is likely to be true. The questioning of the deductive nature of this claim requires that a number of premises be made explicit, including those which embed or embody evaluative claims or norms. The argument here is that it may be better to make a warranted argument than to claim that one's unfalsified claims are simply 'true'.

Review Questions

Describe a situation in science where the transparency of research was questioned, or where the evidence trail or audit trail of a study was considered so important that questioners took the scientist(s) to court.

Thinking of situated knowledge, name five aspects of your own situatedness and how your research might be influenced by those situation-specific conditions.

If you were to offer five criteria for knowledge that might qualify as valid, which would be your best choices?

Looking at typical scientific criteria for valid knowledge, which include offering evidence in support of arguments and having expert peer reviewers, is there a valid-to-nonvalid spectrum? Explain why, or why not.

The reflexive situated scientist might make discoveries, but would other people value those discoveries? Think of a hot topic in social research, such as an ethics-and-medicine topic, or something in the current newspapers. Explain three ways that reflexivity might influence what research methods a person chooses to engage in to study that topic.

Evidence-based arguments: are they really feasible? Break an argument down into premises, intermediate conclusions, and final conclusions. Now express clearly which premises can and cannot be justified using further evidence.

Further Reading

Hughes and Sharrock (1997) review the competing positions on philosophy of science from a sociological standpoint. Kabeer (1994) offers a feminist challenge to orthodox economics, which creates an example showing that the method of falsification of a nomothetic (law-based) account is not a sufficient way to establish the truth of the account. Lawson (2003) offers details of social ontology and defends realism in general against the logical positivism found in some economic models. Olsen (2019) reviews the methodological approach of strategic structuralism, which is also offered at a theoretical level by Dow (2004). Morgan (2013) shows how future trajectories can affect present-day interpretations, showing that most knowledge about explanatory models has both reflexive and normative elements. Reflexivity is also a key theme in Outhwaite's review of philosophy of science (1987).

References

Abou-Ali, Hala, Hesham El-Azony, Heba El-Laithy, Jonathan Haughton & Shahid Khandker (2010) "Evaluating the Impact of Egyptian Social Fund for Development Programmes", *Journal of Development Effectiveness*, 2:4, 521–555, https://doi.org/10.1080/19439342.2010.529926

Archer, Margaret, Roy Bhaskar, Andrew Collier, Tony Lawson, and Alan Norrie (eds.) (1998) *Critical Realism: Essential Readings,* London: Routledge.

Chouliaraki, Lilie, and Norman Fairclough (1999) *Discourse in Late Modernity: Rethinking Critical Discourse Analysis.* Edinburgh: Edinburgh University Press.

Cross, Beth, and Helen Cheyne (2018) "Strength-based approaches: a realist evaluation of implementation in maternity services in Scotland", Journal of Public Health (2018) 26:4, 425–436, https://doi.org/10.1007/s10389-017-0882-4.

Dow, S. (2004) Structured Pluralism*, Journal of Economic Methodology*, 11: 3.

Downward, P. and A. Mearman (2007) "Retroduction as mixed-methods triangulation in economic research: reorienting economics into social science." *Camb. J. Econ.* 31(1): 77–99.

Downward, P., J.H. Finch, et al. (2002) "Critical Realism, Empirical Methods and Inference: A critical discussion." *Cambridge Journal of Economics* 26(4): 481.

Ehlers, T. B., and Main, K. (1998) "Women and the False Promise of Microenterprise", *Gender and Society*, 12:4, August, pp. 424–440.

Fairclough, Norman (1992) *Discourse and Social Change.* Cambridge: Polity.

Fisher, Alec (2001) *Critical Thinking: An Introduction,* Cambridge: Cambridge University Press.

Fisher, Alec (2004) *The Logic of Real Arguments*, Cambridge: Cambridge University Press. (Orig 1988)

Gibbs, Graham R. (2002), *Qualitative Data Analysis: Explorations with NVivo*, Buckingham: Open University Press.

Hill, E. (2001) "Women in the Indian Informal Economy: Collective strategies for work life improvement and development." *Work Employment and Society* 15(3): 443–464.

Hughes, John, and Wes Sharrock (1997) *The Philosophy of Social Research,* 3rd ed., London: Longman.

Kabeer, N. (1994) *Reversed Realities*, Delhi: Kali for Women and London: Verso.

Kelly, Erin L., Phyllis Moen and Eric Tranby (2011) "Changing Workplaces to Reduce Work-Family Conflict: Schedule Control in a White-Collar Organization", *American Sociological Review*, 76:2, 265–290: https://doi.org/10.1177/0003122411400056.

Kvale, Steiner (1996) *InterViews*, London: Sage.

Lawson, T. (2003) *Reorienting Economics,* London and New York: Routledge.

Lewins, Ann, and Silver, Christina (2007) *Using Software in Qualitative Research Step-By-Step Guide*, London: Sage Publications.

Mäenpää T., & Vuori A. (2021). Broaching overweight and obesity at maternity and child health clinics. *International Journal of Nursing Practice*, e12958. https://doi.org/10.1111/ijn.12958, accessed May 2021.

Mearman, A. (2006) "Critical Realism in Economics and Open-Systems Ontology: A critique." *Review of Social Economy* 64(1): 47–75.

Morgan, Jamie (2013) Forward-looking Contrast Explanation, Illustrated using the Great Moderation, *Cambridge Journal of Economics*, 37:4, 737–58.

Olsen, Wendy (2006a) "Pluralism, Poverty and Sharecropping: Cultivating Open-Mindedness in Development Studies", *Journal of Development Studies*, 42:7, pgs. 1130–1157.

Olsen, Wendy (2012) *Data Collection: Key Trends and Methods in Social Research*, London: Sage.

Olsen, Wendy (2019a) "Social Statistics Using Strategic Structuralism and Pluralism", in *Contemporary Philosophy and Social Science: An Interdisciplinary Dialogue,* edited by Michiru Nagatsu and Attilia Ruzzene. London: Bloomsbury Publishing.

Olsen, Wendy, and Jamie Morgan (2005) A Critical Epistemology Of Analytical Statistics: Addressing the sceptical realist, *Journal for the Theory of Social Behaviour*, 35:3, 255–284.

Olsen, Wendy, and Jamie Morgan (2010) "Institutional Change From Within the Informal Sector in Indian Rural Labour Relations", *International Review of Sociology*, 20:3, 535–553, https://doi.org/10.1080/03906701.2010.51190.

Outhwaite, William (1987) *New Philosophies of Social Science: Realism, Hermeneutics and Critical Theory*, London: Macmillan.

Pawson, Ray (2013) *The Science of Evaluation: A Realist Manifesto*. London: Sage.

Pawson, Ray, and Nick Tilley (1997) *Realistic Evaluation*. Sage: London.

Popper, Karl (1963) *Conjectures and Refutations: The Growth of Scientific Knowledge*, London: Routledge.

Porpora, Douglas V. (2015) *Reconstructing Sociology: The Critical Realist Approach*, Cambridge: Cambridge University Press.

Potter, Garry (2000) *The Philosophy of Social Science New Perspectives*, Essex: Pearson Education Limited.

Silge, Julia, and Robinson, David (2017) *Text Mining With R*, Sebastopol: O'Reilly Media,

Smith, Mark (1998) *Social Science in Question*, Sage in association with Open Univ., London.

Touri, M., and N. Koteyko (2014) "Using Corpus Linguistic Software in the Extraction of News Frames: Towards a dynamic process of frame analysis in journalistic texts", *International Journal of Social Research Methodology*, 18:6, 601–616, https://doi.org/10.1080/1364557 9.2014.929878

Van Staveren, Irena (2014) *Economics After the Crisis: An Introduction to Economics from a Pluralist and Global Perspective*, 1st Edition, London: Routledge.

Summary of the Logics and Methods for Systematic Mixed-Methods Research

▶ **Tip** The notion of 'atomism' is associated with strong forms of methodological individualism, which are often applied erroneously. Meanwhile 'holism' reflects that we can appreciate macro entities which evolve over time but are unique. The macro integrates with the micro in ways that make it hard to argue that either is foundational: they interact over time.

This book has focused on the use of quantities in mixed-methods research designs. I stressed the logics used in such situations. There are three areas of 'research design' innovations. First a researcher can use multiple logics of reasoning while doing research, and develop a synthesis; second there are innovative methods such as qualitative comparative analysis and partly-quantified discourse analysis methods; and thirdly, at the stage of analysing and elaborating on themes one applies retroductive thinking to develop warranted arguments. All these innovations are supported by systematic methods, but none of them can simply be assigned to a computer or artificial intelligence, since they require holistic thinking and jumps from micro to meso or macro levels, and seeing linkages between these. In developing warranted concluding arguments a research team normally has to make judicious decisions about which mechanisms or interpretations to put stress upon.

In bringing these three themes together, an approach known as holism comes to the front. Holism is the notion that wholes exist and that they influence the little partial entities within them, and each other. Holism itself is related to causal holism, in which a causal factor might be a large-scale, macro or meso factor and potentially influence, or shape, many micro entities (class affects status of individuals, for example). Holism also brings to the fore that the researcher is a part of the scene, not out of it. Therefore a summary of logics in research explains explicitly how we use holism to good purposes.

Holism is also related to the nature-nurture debate and to the sociology debate over human's animal nature versus our 'second nature'. In the nature-nurture debate, human bodies are often taken to be fundamentally, structurally identical across the

© The Author(s), under exclusive license to Springer Nature Switzerland AG 2022
W. Olsen, *Systematic Mixed-Methods Research for Social Scientists*,
https://doi.org/10.1007/978-3-030-93148-3_10

world (on the 'nature' side) even though evolution and lifestyles do amend each person's body towards locally specific patterns, such as tall nomads. The 'nurture' side of this debate puts more stress upon the human lived experience through a human life, which can be seen as making each person unique. On the nature side, biology and chemistry may be stressed, whilst on the nurture side there is stress on the social communication, learning, and behaviour aspects. Holism is not the idea that culture dominates. Ontological holism means that we respect the large-scale entities, whether we look at biology, chemistry, or the social. Even in biology and other natural sciences, there is stress on emergent properties of structured entities, such as bacteria, and thus atomism does not dominate there. A balance of considerations allows your project to consider both natural and physical aspects, and social/cultural aspects of any given topic. In the sociology literature, interesting discussions took place about humans' 'Second Nature' (Bauman 2010). The very concept of 'nature' is only and always defined within the context of human cultures, and thus is in a dialectical and dynamic relationship with human society. Therefore, the researcher who works in an area affected by 'nature-nurture' debates has much freedom to try to shift the terrain towards progressive social framings of a topic. Those working on topics in engineering, forestry, fishing, and others related to 'natural' science can bear this in mind. This chapter will explore and summarise the types of reasoning that can be used.

Given the depth-ontology starting point, Chaps. 1, 2, 3, and 4 have discussed five logics of reasoning if counted separately: induction, deduction, retroduction, synthesis, and logical linking. We can use them in sequence and in iterative steps in a flexible order that allows feedback. This process then opens up revised 'arena' questions, questions about the scope and range of the research question. Many scholars have already made this point that the findings may lead the research team back to re-assessing the fundamental theoretical starting point. In this book it is often explicit (and also explained in Olsen 2019b). Therefore a reformulation of the 'research question' can occur.

10.1 Induction

The first type of logic was induction, a move from particulars to general statements. Induction can underpin some kinds of claim that appear 'factual'. It can also underpin abstract claims (Byrne 2005). An inductive abstract statement must rest at some point in its reasoning upon evidence or upon the real-world observations made by a person. Induction can be seen as quite narrow and particular, but it is also possible to use induction to move to general statements (see Sayer 1992 Chapters 1–2). Inductive conclusions can be challenged on two grounds—first how generalisable are the claims across space, or time, or social scenes? And second, how far did a presumed narrative couch the findings in a given language, and was this initially presumed narrative taken for granted? It was perhaps not re-assessed for its usefulness and fit, making it subject to contestation. These challenges lead research teams to supplement their inductive methods.

There is a large grey area of contested and contestable claims which have some claim to validity through inductive reasoning. I argued that evidence could be used to falsify someone else's inductive claim. However, again, the points made by each side in an argument can all be subject to further questioning or even subject to attack (see Chap. 8). It was stressed that simple forms of induction from data to findings can be enhanced through further elaboration, but the elaboration stage involves social theorising and may not be simple 'induction' because we invoke reasoning that is more complex.

10.2 Deduction

A second logic was deduction. There are two forms of deduction in the typical mixed-methods context. First, there is usually a deductivist approach in mathematical derivations. These derivations (such as proofs, lemmas, and corollaries) do not constitute a high proportion of the research design in any social science project, whether mixed methods or not. I took a strong view against the nomothetic hypothesis-testing school of thought if and when it is linked to the claim that the conclusion of a deductive argument is a social fact. The concept of a fact used in this book differs somewhat from that found in Thomann and Maggetti (2017). In this book, a fact is a claim about reality, while in Thomann and Maggetti (2017) the facts are defined as claims whose validity or invalidity is a matter of objective assessment. I have argued that with a depth ontology the researcher is 'in' the scene and not 'outside' the scene, and that there is a grey area where claims may be contested both on ontic and subjective grounds. My first point is ontological; Thomann and Magetti has tended to collapse the real, the actual and the empirical into a single layer, and thus their claim is more epistemological. Another way to distinguish this book's orientation from the one expressed by Thomann and Magetti, which is commonly held as a view of truth, is that in this book the truth status of a claim depends on reference to the world and not just on people's assessments. In other words there is a possibility that humans might err when making supposedly 'objective' assessments, precisely because our assessments also have a subjective dimension.

We have made sense of this by discussing the overlap areas of the subjective (socially constructed) and the objective worlds. Olsen and Morgan (2005) presented arguments in favour of realising that facts and ficts are different, and that both can be contested. We argued (ibid., 2005) that the truth of facts when they are buttressed by analytical statistics is not just an epistemological problem, but an issue that raises ontological questions. With those claims we call facts, we make recourse to reality and hence to evidence. Whether a fact is really true does not depend upon the manner of its assessment. This may be unfamiliar territory for some people. I have tried to summarise a reality approach to truth here but more could be said to flesh it out. The way we test factual claims for their validity also varies, depending on what kind of claim is being made.

This book has dealt carefully with the potentially useful, attractive practice of 'testing a hypothesis' which is a second form of deduction. This routine is typically also rooted in the nomothetic idea that what is true for one group of cases may be tested upon another group of similar cases. (We can also test holistic hypotheses. These involve small-N or a sample of just one item. But more often, deduction with quantitative data has been conducted with regard to atomistic hypotheses about groups of cases.) The actual testing of a hypothesis can lead us to new discoveries. Testing can aid in generating useful conversations about policy or norms, or about the world itself. This testing takes a typically deductive format: review the literature to derive theories [or explanations], formulate testable hypotheses from these, gather data, test the hypotheses, and decide which are falsified. In sum, testing can be seen as a deductive activity.

There were two reasons for questioning social facts, parallel to those which can be used to challenge inductive reasoning: The first is to ask how generalizable would the 'law' (*nomos*) or the hypothesis be? I particularly challenged individualism and atomism for an apparently excessive generality throughout this book. The risk of spreading a truth about a sample to a larger group of entities, beyond that sample's population, is present in hypothesis testing. This challenge can be easily addressed using careful research design and clear *ex post* writing. Another way of putting this is that inference from sample to population must carefully set boundaries on external validity. Claims may not be valid beyond the sample data and beyond that data's time-period. Secondly, I expressed concerns about unquestioned initial grounding narratives. It is possible for the research team's grand narratives or discourse habits to lie quiet, unaddressed, and uncritiqued; and yet to exhibit themselves in key ways, because there are discourses that underpin power and authority in any society. Research write-ups are not immune to the suffusive nature of social power. Whilst power has many meanings (power-over, power-with, capacity-to, and the ability to shift an agenda), in all these cases it is generally the case that discourses underpin power and power suffuses social data and hence affects claims about society. I gave examples from banking where bank loan forms funnelled a gendered attribution of assets to males instead of females (Chap. 8).

However there is still a place for deductive reasoning in the pantheon of methods. It is used to get from one pattern of data or algebra to another. Deductive reasoning is not the same as 'the scientific method'. Any textbook that equates these two is going to run into contradictions and difficulties. Deductive reasoning encompasses a much wider range of modes of reasoning than mere scientific hypothesis-testing. As examples, I have offered multilevel modelling shrinkage estimates (Chap. 4), QCA Boolean truth-table reduction (Chap. 6), and the analysis of coded qualitative data 'nodes' (Chap. 8). Deductions sit within more complex argument types and indeed, argumentative styles. At the same time, scientific methods can include non-deductive steps. So there is no rebuttal of whether one's research is scientific if someone claims that it lacks deductive logical moves.

In the qualitative chapter of this book, I explored a few ways to unpick the given discourses, make them explicit, name them, and analyse them. In this process the researcher is within the research scene, not outside of it. This idea of the embedded researcher underpins the 'situated knowledge' approach to science, found in Chaps. 2, 3, and 8. It is a solid foundation for research. I assume that research occurs in a context of social dialogue and is not conducted by a person with a 'God's-Eye View'. The foundations of research are a matter for discussion and are not fixed.

10.3   Retroduction

During a project, we revisit the scene of the research or the existing dataset, and we explore these resources for ideas that help explain what we are seeing. This is known as retroduction, or asking 'why?' in a self-guided way. With retroduction, we re-state key points of knowledge that are taken as given or which may be present as an absence in the research data. Noticing an 'absence' is the more problematic and surprising move. It requires that we have existing knowledge—or that we can widen our knowledge—to realise that this absence is a key one out of all the other things that are absent. Substantively, we also can consider why the data look a particular way and generate possible themes to explain why the data have those patterns. Here, retroduction is like explaining the findings. I distinguish the closed form of retro-duction, when we limit the explanation to existing data, from the open forms which extend our thinking to anything that is among the conditions of possibility of the observed patterns and data. Closed retroduction uses more of the available data. Open retroduction involves going back to get new data, or using fresh perspectives to revisit the field scene.

Retroduction sits well with either induction or deductive reasoning but is not the same as either. According to Danermark et al. (2002), retroduction fits with a feed-back loop or iterative approach to research design, since numerous re-attempts to explain the situation can be deliberately carried out.

I was keen to enhance retroduction by arguing for three new forms of discovery. Here I summarise:

- There are discoveries of key meso-level entitles which affect outcomes across a wide swathe of the data, found either through multi-level modelling (Chap. 4) or through discourse analysis (Chap. 8).
- There is the realisation that some profound historically unique condition is act-ing as a key mechanism to produce outcomes in the data. For example to discover and then argue that capitalism has, as an essence, a drive to productivity or a drive to accumulation, would be a form of retroduction. Authors are then ready to engage in theoretical debates about the articulation of different forms and systems of capitalism with other social entities and with nature.
- Lastly, there is the realisation that there is macro-level diversity with small 'N' and that it matters. Hospitals or universities in two regions operating under dif-ferent taxation regimes within one country might be an example. At issue is not just that the country is federal nor that the tax regimes are what they are, the key discovery is, **to what do** the **two** taxation regimes make a difference? Those accustomed to survey data and regression methods may need to practice macro thinking. One can practice: (a) drawing upon the essential features of the macro entities to develop an argument about what their effects are; (b) drawing upon

historical evidence to discern the roots of the differentiation of tax regimes; and (c) convincing audiences that it is not just a matter of dummy variables but profoundly the social-specific features of the hospitals (or universities or tax regimes) in **each** region.

10.4   Synthesis

The fourth logic is synthesis where we combine statements from a range of elements in a study. The statement types include premises, which may be assertions or assumptions; data and induction; deductive logic perhaps; multiple stakeholder voices; intermediate conclusions; and so on. Synthesising these is a way to reach some conclusions, focused perhaps on your original research question, while also being ready and willing to draw conclusions that are surprising to you or your team. I argued that as a career progresses, individual researchers move from team to team, and from project to project. Synthesis can happen at random or in a concerted and documented fashion. Write-ups and presentations contain a synthesis. A dialogue ensues. There is no final conclusion.

These realist methods strongly contrast with the philosophical 'syllogism', often taught in textbooks. A syllogism is a form of reasoning from Premises to Conclusions that makes the intervening logic explicit. The syllogism is a way to place a series of statements in relation to each other. The contrast is that realist statements make reference to underlying entities, not merely to the statements themselves (Olsen 2019a). 'Deep linkage' and in-depth interpretation take us beyond syllogistic reasoning. Perhaps a third form of logic is to draw deep linkages between different types of evidence, and in so doing, to come to conclusions about what is happening at a deeper level in a society.

For Bhaskar (1989), a critical realist philosopher, one key element of scientific realism that was missing in the 'hypothesis-testing' approach was meta-critique (critique of theoretical starting points). I concur with that point. If we look at women's mental health (Chap. 1), there are social class issues, labour-market issues, and post-colonial issues. If we look at homelessness and the anti-homelessness social movements (Chap. 2), there are again social class issues and there are issues around migration and ethnicity. Thus, in carrying out a meta-critique, a research team often has to deal with wider issues than they originally expected to look at. Mixed-methods research may draw a project away from one specialism and into contact with others on a broader stage.

There is a risk of superficiality in induction and deduction. Induction might suffer from assuming all the evidence is composed of true facts; deduction might suffer from not referring to latent entities. These were points that Bhaskar (1989) and Sayer (1992, 2000) made from a realist point of view. When we bring together multiple logics in a complex synergy, we can avoid some of these more obvious weaknesses.

Once the iterated feedback looping starts, we may find errors and mistakes in the data. We may conclude, after doing deep linkage with qualitative and quantitative evidence, that some of the earlier claims were misleading and could lead the reader up a side alley. We develop a new focus and fresh insights. This move generates significant, original, innovative research. Mixing methods is helpful when aiming for this stage.

10.5   Recognising Relevant Irreducible Phenomena (Holism)

Data reduction was described in Chaps. 5 and 6, but data reduction was not a move in logic. It was mainly a handy way to simplify a data table. One data reduction method for survey data was latent variable analysis, which helped us create a scale. The data-reduction method used in QCA and fuzzy-set analysis is typically a Boolean simplification of a true statement about a situation (Chap. 6). We might move from fifty individual pathways to reach condition Y and then summarise them as ten main pathway types to reach Y. Data reduction via Boolean methods might even lead to a further simplification. (A is necessary for Y but three pathways are each sufficient for Y, given A.) In QCA, another form of reduction also occurs. I call it fuzzy intersection. If we look at a configuration containing characteristics R, F, and A then we are taking $R \cap F \cap A$ which means the intersection of these three. We would call this a new fuzzy set, and give it the name X. It has its own range of measurements: It will tend to be lower than R or F or A, because it takes the minimum value of R, F, and A for each case. Data reduction that simplifies a data set leads to straightforward hypothesis tests, which may be powerful and encompassing rather than simply leading us away from true detail. Thus, the data-reduction methods listed here enable us to develop arguments that have a strong purchase on the world, rooted in data from a concretely identified part of the world in space and time.

Thirdly, in Chap. 8, data transformation also gave a reductive glimpse of a large textual database. Thus there were three main methods of 'data reduction' in the book.

- Latent variables can enable survey dataset reduction
- Truth table reduction can simplify causal pathway representations
- And discourse codes can indicate key aspects of discursive patterns

If data reduction is considered desirable, we can also ask: Is there any phenomenon which is *irreducible?* If so, is this part of its essential nature, or is it just a question of our subjective choice to see it as an irreducible whole? A simple example is that the class structure is a whole set of classes and class relations. If a class structure has features which cannot be reduced to these constituent elements, then it is, as a whole, a unique and special thing. We may say a class structure is irreducible.

The fundamental logic of the realist argument is that if something is irreducible, then calling it by a given name helps to draw upon and highlight the key essential facets of that 'thing' which make it so. We do not argue that naming something exhausts the nature of that thing. Naming something is also not going to solve social problems. Naming things is, however, a useful step in a scientific dialogue. One example I gave was listing some key differences between social norms and individual attitudes. By naming attitudes we enable ourselves to think in a proper way of deviations from social norms, and by naming mezzo-rules—those rules in society which are not entirely followed (Olsen and Morgan 2010)—we enable ourselves to understand better the overall dialectics of a complex social situation. So naming a 'whole', respecting its attributes, is a useful skill. Quantitative social scientists will want to practice not only regression methods but wider, deeper, and more interdisciplinary forms of argumentation that use wholes as objects in the world. We cannot study only the multiplicity of individual units.

In all three of the data reduction methods, which are compatible with each other, the researchers can use either closed retroduction to find out what is the strongest explanatory factor in a given set of data, or open retroduction to derive an impulse to exit the dataset, refresh and expand it, and later go back to drawing conclusions. In this way, data reduction is not the end of the process, it is one part of the research process. I challenged the critics of QCA, notably those who want the data reduction to be 'complete' and 'correct' (Baumgartner and Thiem 2017), or those who are apparently atomistic in ontology (Lucas and Szatrowski 2014). The excessively atomistic (non-holist) approaches are widely known as methodological individualist. I offered a challenge to individualism and atomism.

The presence of holism, and the existence of key entities as wholes, is likely to make the research team realise the importance of going back to the field (open retroduction). Ragin in particular has encouraged both of these as good ways to do sociology and political research (1987, 2000, 2008). Another widely respected author, Charmaz (2014) also advises that revisiting the field will help in a concluding analysis. In this book, I have argued that holistic objects and irreducibility are related. Yet those who do not call realism by its name have perhaps avoided having the arguments that will push more researchers forward to the following important realisations:

- Holism: The embeddedness of cases in wholes, whether macro or meso-institutional factors 'above' the case level. Therefore, seeking to know more about whole entities is valuable.
- Openness: Many systems in society are open, and not closed. Obvious examples are the food system, the climate system, and the class system. Open systems change from within.
- Non-reduction: To explore why, we often resort to examining wider aspects like historical context, regulatory systems, corrupt environments, politics, or social norms of sub-groups. (Also called the irreducibility of wholes.) These are irreducible to the micro-level cases.

10.6   Logical Linkage

The fifth mode of logic was to piece together an analysis using elements of all the above modes of reasoning: linking logics. A wide range of mixtures is possible. I described the various exemplars in Chaps. 2 and 4 to show what has been done in studies that typically use quantitative methods along with—or perhaps whilst in need of—mixed methods.

▶ **Tip** In your research, you may come across theories or claims that seem wholly wrong to you. The task is not to brand them as wrong, or as fictive or worthless, but rather to respectfully unpick their logical arguments and critically assess their evidence, and then point out which premise is problematic.

10.7 Conclusion

Once we have absorbed all the five logics above, we realise that the conclusion of a study is its claims, not just a regression results table or a reduced-form equation or a model. Both statistics and qualitative comparative analysis (QCA) can be improved by recognising this point. It would be wrong to criticise QCA for having the 'wrong pathways' in its reduced-form equation, since the issue is not about the reduced truth table but what **claims** were made about the world. It is also very wise to explore the kinds of mistakes that may take place using the regression, fuzzy-set or crisp-set QCA frameworks, or discourse analysis.

In conclusion, researchers should take on board all five logics and learn to use them flexibly. Now I turn to summarising the methodological claims.

In this book I described systematic mixed-methods research (SMMR) as those studies which contain a systematised data table but still have depth thinking or a depth ontology, and use multiple methods to deal with the different layers.

I also described how action research could use realist methods of critique and retroduction. This meant stakeholder voices could be brought to the fore in research. Many studies of this kind exist. However it is not necessary to call our analysis of them strong social constructivism. We can use a weak, combined form of social constructionist methods. We must not ignore the various insights we can get from the realist approach and from retroduction to social-structural and institutional causal mechanisms. I call the discourse analysis done by realists weak social constructivism instead of strong social constructivism. I stress the realism aspect, and the practice of naming via essential features. We can be both confident and self-reflexive at the same time. Deep linkages, notably naming discourses and creating latent variables, encourages us to use many logical steps that do not fit in with the strong social constructivist approach. Overall I took a methodological pluralist approach.

In epistemological terms, using statistical measures of good fit helped with finalising the data reduction, and yet it did not lead to 'conclusions'. The step of drawing conclusions is a step carried out by humans. This step is necessarily couched in language that you carefully hone, using not only the statistical model but several 'Arena Stage' matters such as which disciplines you cover, what is the leading theory, what theories you criticise or omit, and which levels of society to handle explicitly. We cannot create a black-box computer package to carry out mixed-methods research.

I have suggested that retroduction works well when a variety of data types are available. I could not build a strong case for a project using only statistical data unless it was, in some sense, a very small project. It might be small in geographic scope or in terms of work-time inputs. Survey data are, themselves, always shaped around deep (either explicit or implicit) conceptual frameworks such as discourses of measurement and the local lay conceptualisation of entities. As a result, theoretical frameworks are always mingled with numeric measures in statistical research about the world. Therefore even a 'secondary data analysis project' could benefit from examining the multiple layers of meaning that we explored in Chap. 8. The social statistician or a statistical expert in the sciences is well served by spending time learning how to carry out qualitative analysis. I stressed that meso and macro entities are part of the qualitative scene. Researchers can analyse not only micro-meanings in discourses but also the meta-meanings and the real impact of things, processes, and entities, which underpin or surround or influence the survey data. Having skills with multiple data types, and mixing methods, is even more important if you are using unstructured textual data, as found in some data-science projects.

Generating New Research Ideas

The five logics described earlier and the broadly meta-critical approach to theory can also be used to generate new research hypotheses, new problems, and new ways of addressing long-standing questions in social sciences. The book has integrated these approaches.

Review Questions
Name two differences between induction and retroduction, giving an example.

Can the standard nomothetic deductive approach to hypothesis testing be used with a macro object? State reasons why or why not.

Thinking of deep linkage and cross-level interactions, what are three advantages of mixed-methods research that includes both qualitative techniques of data analysis and quantitative statistical analysis?

What is a holistic logic, and what are two differences between holism and atomism when they are used in a study of, for instance, marriage?

Further Reading
Flick (1992) introduces the two possible ways to approach epistemology. First you could assume all evidence is correct and well set out, but second you might begin to question the underlying frameworks of reference which set up a discourse within which key evidence is couched. Plano Clark and Creswell (2008) offer a variety of discussions of mixed methods which touch on the validity and science issues. Lastly, Archer et al. (1998) is a useful compendium and reference work on realism. It covers mainly how theory and concepts lead towards debates in science. It is not limited

to social science. For a social-science focus, we can use the following four text-books on philosophy of science as reference works: Bhaskar (1979), Blaikie (1993), Potter (2000), and Outhwaite (1987).

References

Archer, Margaret, Roy Bhaskar, Andrew Collier, Tony Lawson, and Alan Norrie (eds.) (1998) *Critical Realism: Essential Readings,* London: Routledge.

Bauman, Zygmunt (2010), *Towards a Critical Sociology (Routledge Revivals): An Essay on Commonsense and Imagination,* London: Routledge (originally 1976).

Baumgartner, Michael, and Alrik Thiem (2017) "Often Trusted but Never (Properly) Tested: Evaluating Qualitative Comparative Analysis", *Sociological Methods & Research,* 49:2, 279–311. https://doi.org/10.1177/0049124117701487.

Bhaskar, Roy (1979) *The Possibility of Naturalism: A Philosophical Critique of the Contemporary Human Sciences.* Brighton: Harvester Press.

Bhaskar, Roy (1989), *Reclaiming Reality: A critical introduction to contemporary philosophy,* London: Verso.

Blaikie, Norman (1993) *Approaches to Social Enquiry.* Cambridge: Polity.

Byrne, D. (2005) "Complexity, Configuration and Cases", *Theory, Culture and Society* 22(10): 95–111.

Charmaz, Kathy (2014) *Constructing Grounded Theory,* 2nd ed., London: Sage.

Danermark, Berth, Mats Ekstrom, Liselotte Jakobsen, and Jan Ch. Karlsson, (2002; 1st published 1997 in Swedish language) *Explaining Society: Critical Realism in the Social Sciences,* London: Routledge.

Flick, Uwe (1992) "Triangulation Revisited: Strategy of Validation or Alternative?" *Journal for the Theory of Social Behaviour* 22(2): 169–197.

Lucas, Samuel R., and Alisa Szatrowski (2014) "Qualitative Comparative Analysis in Critical Perspective", *Sociological Methodology* 44:1–79.

Olsen, Wendy (2019a) "Social Statistics Using Strategic Structuralism and Pluralism", in *Contemporary Philosophy and Social Science: An Interdisciplinary Dialogue,* edited by Michiru Nagatsu and Attilia Ruzzene. London: Bloomsbury Publishing.

Olsen, Wendy (2019b) "Bridging to Action Requires Mixed Methods, Not Only Randomised Control Trials", *European Journal of Development Research,* 31:2, 139–162, https://link.springer.com/article/10.1057/s41287-019-00201-x.

Olsen, Wendy, and Jamie Morgan (2005) "A Critical Epistemology Of Analytical Statistics: Addressing the sceptical realist", *Journal for the Theory of Social Behaviour,* 35:3, 255–284.

Olsen, Wendy, and Jamie Morgan (2010) "Institutional Change From Within the Informal Sector in Indian Rural Labour Relations", *International Review of Sociology,* 20:3, 535–553, https://doi.org/10.1080/03906701.2010.51190.

Outhwaite, William (1987) *New Philosophies of Social Science: Realism, Hermeneutics and Critical Theory,* London: Macmillan.

Plano Clark, V. L. and J. W. Creswell, eds. (2008) *The Mixed Methods Reader,* London: Sage.

Potter, Garry (2000) *The Philosophy of Social Science New Perspectives,* Essex: Pearson Education Limited.

Ragin, C. C. (1987) *The Comparative Method: Moving beyond qualitative and quantitative strategies.* Berkeley; Los Angeles; London: University of California Press.

Ragin, C. C. (2000) *Fuzzy-Set Social Science.* Chicago; London: University of Chicago Press.

Ragin, Charles (2008) *Redesigning Social Enquiry: Fuzzy Sets and Beyond,* Chicago: University of Chicago Press.

Sayer, Andrew (1992 (orig. 1984)) *Method in Social Science: A Realist Approach.* London, Routledge.

Sayer, Andrew (2000) *Realism and Social Science.* London: Sage.

Thomann, E. and M. Maggetti (2017) "Designing Research with Qualitative Comparative Analysis (QCA): Approaches, Challenges, and Tools". *Sociological Methods & Research,* 20:10, https://doi.org/10.1177/0049124117729700

Glossary

11

Action research[1] A method of involving stakeholders in carrying out change in a bounded social scene, such as a regulated marketplace or a firm. The method includes an initial review, task planning and task allocation, at least one but preferably two or more cycles of feedback, and revision of the tasks and perhaps changes in organisational structures, regulations, or norms. A beach would be a potential scene of action research, bringing together stakeholders who may not have previously met each other.

Association This is a word we can use for correlation when the variables are not continuous in their level of measurement. One measure of the association of a series of binary variables for example is their correlation. But the correlation of a bunch of ordinal variables using Spearman's correlation coefficient is often referred to as a measure of their degree of **association**, rather than correlation per se. The ordinal variables do not have linear relationships with each other but may be monotonically related, or related in a positive way, via their ranks. The Spearman's measure uses the rank information rather than the cardinal information. To be associated thus means to co-occur with something.

Boolean An algebra or a calculative logic that has four main features: True/false (T/F) or yes/no; the operators 'Or' and 'And', also known as disjunction and conjunction, or union and intersection; and finally the negation 'Not'. 'Not' also reflects absence of something. In an example we could say: 'Religious heritages in India rest upon a mixture of lineage and tribe, along with beliefs and practices, represented as each case having T/F for each of the combination represented by one algebraic expression, L&T&B&P, in which many-valued multinomials

---

[1] **Tip**
Draw a Venn diagram showing which disciplines you are working within. The circles show how they overlap. The use of language from this glossary broadly encompasses most discipline areas, but it is important to be sensitive to discipline-specific usages. You may stipulate two alternative definitions for one term to make your work more usable to your key discipline-specific audiences (e.g. medicine, business, forestry, etc.)

© The Author(s), under exclusive license to Springer Nature Switzerland AG 2022
W. Olsen, *Systematic Mixed-Methods Research for Social Scientists*,
https://doi.org/10.1007/978-3-030-93148-3_11

are used to represent possibilities: L for lineage, T for tribe or clan, B for belief system, and P for religious practices. In such a case, L&T&B&P takes up to 24 combinations.' A Boolean algebra does not have to be restricted to 0/1 binary variates, because we can allow for multivalued options like L. Usually in Boolean algebra the Not-X is mutually exclusive with the X, for each X. Therefore if we wish for belief systems to overlap, we have to allow for each one as a binary variable, rather than B as a multinomial. Thus: H=Hindu belief system (0/1), M=Muslim belief system (0/1), and so on, will allow the coexistence of two or more beliefs in one person.

Calibration A method of transforming the measurement of a variable or a variate, usually to create a rank-preserving rescaled version of that variable.

Cause and effect An enumeration of two factors, one having the capacity or tendency to create the conditions for the occurrence or existence of the other. The cause may pre-exist or co-exist with the thing that we consider as its effect.

Conditions of possibility Once the evidence has been obtained in a project, the social researchers may enquire about the historical, social, political, and other conditions which have set the scene for this particular project and its measurements or other evidence. To ask about the conditions of possibility of the outcome in a regression, we make a retroductive enquiry. So the conditions of possibility of asking these questions include having a narrative of a particular type which is free of certain restrictions which are commonly found in the lay discourses of the scene which is being studied.

Correspondence analysis A method of analysing the 'mass' of a group of variables, including any level of measurement such as nominal, ordinal, and continuous variables, leading typically to 2–3 axes of differentiation and the plotting of supplementary variables into the space of the first two axes. Closely related to exploratory factor analysis. Also called multiple correspondence analysis (MCA).

Counterfactual For an explanatory claim such as X is part of the causal underpinnings of outcome Y, the counterfactual is what would have happened if not X. Thus in general, counterfactuals are reckonings of what could or would have happened in the absence of an interrupting reality. In treatment situations, one counterfactual is what would the effect of T have been among the untreated? Another is what would the effect of T have been among those who had Not-X, whilst we have data only on those who had X. Counterfactual claims are not artificial, because they describe a reality which had both actual historical existence and some essential features. The counterfactual helps us perceive what the interaction of those essential features is, or would be.

Discourse A discourse is a set of rules guiding the use of communicative acts, notably language or body language or other symbols, which in a specific historical and social situation have recognised meanings and implications.

Endogeneity In the regression context, endogeneity is the presence of strongly correlated variables which are in the group of 'independent' variables.

Epistemology An approach to the validity of knowledge that unearths or discusses the grounding of knowledge in either principles, or practices. Many epistemologies exist, and most can be critiqued. Three commonly mentioned aspects of modern scientific epistemology are validity, reliability, and replicability. Feminists and humanists argue that epistemological claims imply values, and that science is not value-free.

Exogenous In the regression context, a new variable which is unrelated to all the other variables.

Information content of a dataset The information held by a single random variable can be summarised using the inverse of the Shannon Entropy Measure or one of its variants. Therefore the information held by a binary variate can also be represented easily in algebraic terms (Frank and Shafie, 2016). For instance, a uniformly distributed continuous variable has very little information, whereas a variable with a tight distribution has more information. Among fuzzy-set measures which are bounded by 0 and 1, those near 0 and those near 1 have the most information, while those which are set at 0.5 have little information. We can think of informativeness both on a group of cases, with values that vary or do not vary; and also on the expectation of the variable if it is a random variable. Thus information content of a dataset has both an a priori value (which we can reckon if we have some knowledge about the dataset's features) and an ex post value which arises from the actual data. The Shannon measures allow researchers to compare on an explicit, numerical scale the relative informativeness of different combinations of data.

Intertextuality Fairclough (2001) denoted intertextuality as the co-appearance of two or more discourses in one text. He noted that we expect to hear and see people mixing discourses because life is a complex interplay of many realms, each of which has discursive rules, and because of overlapping conflicting rules of discourse. Intertextuality is when we see a metaphor or idiom from one discourse used with irony or sarcasm in a statement from another discourse, for example. To be more specific, in a social media discussion one may say 'I am blocking you' or one may say 'If you do not stop cursing, I will block you', and the second one is intertextual. It reflects the social media discourse pattern of threatening to block someone, but it also names the social rule that cursing is undesirable, and thus draws upon a much larger, prior, social discourse that is not dependent upon social media discourse. Intertextuality is also a way for creativity, resistance, art, and novelty to play out in human creations.

Keyness The prevalence of a word within a body of text can be called its keyness. The body of text, or 'corpus', may be rather large and non-typical so we examine the relative prevalence of this word in this text, compared with the prevalence of the word in a much larger or broader corpus. This relative keyness can be measured using an odds ratio: the ratio of the odds of it appearing in the first corpus divided by the odds of it appearing in the larger, broader, and even bigger corpus. Phrases could be used instead of words if preferred.

Layers of meaning In social life, discourses utilise words, phrases, and gestures or iconic images which carry communicative meanings, but these are multiple; therefore layers of meaning refer to direct meaning within a phrase, then a higher layer of normative overtone or undertones of meaning (e.g. carrying disapproval along with saying 'So you live with her?!'), and other layers which may invoke political commitments, religious discourses, and other patterns in society that bear meaning. A simple choice of a colour in an advertisement may bear a layer of meaning. We do not say it 'is that layer of meaning', because that would be ontologically badly framed. We can say the symbolic meaning of the colour is a layer of meaning, in addition to other perhaps more obvious or explicit layers. An example in banking is the use of pink colour in Western societies to encourage women to create savings or pension accounts. Where pink reflects a 'meaning related to femininity', the colour may bear some notion around the concept of a feminine capacity to care for herself.

Meta-critique An analysis of someone's claims by questioning not only the claims made about the world, but also the fundamental framings and the discourse chosen. For example one might question whether the banking system analysis was appropriate in its coverage if it ignores the whole shadow-banking economy. One would question the labour-market analysis if it left out domestic work and social reproduction. In carrying out a meta-critique we ask why the omission or error occurred. It may not be deliberate *per se*, but rather a reflection of social trends. By discussing these trends, the research is raised from atomistic to holistic issues; from micro to macro issues; from perhaps a non-historical approach to a more historically grounded approach; and from perhaps a non-normative approach to one which is explicit about underlying and stated norms. It is possible, using meta-critique, to identify hypocrisy that is not intended by a speaker. Touri and Koteyko (2014) illustrate by displaying neoliberal contradictions in discourses of public speeches.

Methodological pluralism This complex topic refers to using a mixture of methods, with the qualitative part being fundamentally rich and inclusive; thus Roth (1987) on merging qualitative and quantitative data promotes methodological pluralism. There is also a hint of theoretical pluralism in the literature around this topic. Finally, methodological pluralism can also mean mixing logics, such as combining a falsification step in a project—which satisfies some more empirically minded scholars—with a phenomenological or critical realist step. Model 1: $Y = a + bX + cZ$ where X, Y, and Z are random variates for cases 1 to n. Model 2: $Y = a + bX + cZ + dM$ where M is another random variate on the same cases.

Modernity A social system which has multiple interlinked applications of complex systems of management, such as air conditioning, architectural drawings, and water systems, associated with a desire among elites for systematisation. For example a modern school would have electronic communications between teachers and management, harmonised classroom arrangements, and a clearly expressed grading system. **Late modernity** is a combination of complex systems of harmonised management along with cultural mixing due to globalisation.

Multi-level model A regression model with variables for the individual cases and other variables reflecting causality in the aggregate with things at a higher level. A two-level model will have the level one cases nested in level two. With three or more levels, non-nested situations are possible. An example would be firms and regions with individuals as level one. Each firm may be in multiple regions, while each region has multiple firms. Multilevel modelling can handle both nested and non-nested levels with about two to four levels being typical in a regression model.

Nomothetic A systematic way to do science by seeking to work out the laws of the social and natural scene.

Ontology Social ontology is the study of what exists, taking into account the great complexity of interlinkages. The act of ontic exploration is involved in both knowing and then expressing what things are really like: types, characteristics, essences, relationships, and causal mechanisms for example play a key role in social ontology.

Open retroduction The process of asking why, and searching for new evidence, in order to improve our understanding of how something works or why something happens. We are willing to re-define the ontological depth of the analysis, and we may change the geographic or social scope of a study if necessary.

Optimal model In statistics, the models are typically linear or non-linear, or multi-stage generalised models, and we can compare models. We may compare them based on the likelihood of each model (which has a mathematical expression), the probability of each model (using Bayesian reasoning), the conditional probability of each model (given the data), or which one minimised the squared errors of the cases, that is the residuals. In Boolean algebra, an optimal model might be arrived at using the same or different logic. First a maximisation approach could be used if the Boolean model were re-expressed in ways that provide a distance measure, residuals, or probabilities of models. Second, the Boolean model would be optimised by its accurate representation of reality, but this may be more of a theoretical and less of a mathematical approach. Thirdly, the Boolean model that fits best might have the highest consistency with a particular initial hypothesis, such as: X is sufficient for Y. Another hypothesis is X is necessary and sufficient for Y which requires different distance measures.

Ordinal A measure with multiple ranked levels is considered ordinal. Beginning with a binary, which have values 0 and 1 (0 meaning absent and 1 meaning present), the ordinal measure can take a fuzzy-set membership score value along the numberline from 0 to 1, or ordinal measures can be unique values on the whole real number scale such as -7, 1, 10, 14. Ordinal variates hold the level of the rank of each case. Therefore we could have ten cases valued at: -7, -7, -7, 1, 1, 10, 14, 14, 14, 14. There are four ordinal levels in this example. All continuous and ratio variables have the 'nature' of being ordinal as well as continuous. However, we never refer to binary variables as being ordinal because the 0/1 comparison is simpler.

Realism Realism is a philosophical approach which argues that some entities pre-exist and influence the social researchers when we try to carry out research. Realists also allow for things that are 'macro' entities beyond any one individual's control. A macro or meso thing can be influenced by people, but is going to resist change because of its own inherent or emergent properties. An example of an inherent property is that a speculative housing-price bubble has both a tendency to continue growing and to burst. Those trying to influence house prices cannot easily predict or control when the bubble will burst.

Reflexive The reflexive methodologist allows for their own introspection, and dialogues among research participants, as a fundamental part of the learning that goes on during social research. In this context, reflexive means reflecting upon one's own approaches, one's language and body responses, and others' language, and moving towards new images of the scene; also known in Mills' writings as '*The Sociological Imagination*' (Mills, 1959).

Retroduction Retroduction refers to drawing inferences about the best answer to a key question. One works backwards from what is known, or from the evidence, asking questions to force oneself to discover or create explanations. Three levels of retroduction are: asking why the data have a certain pattern (a question about reality); asking why some data are framed in a particular way, for example scientists using a particular theory (a question about social constructions); and thirdly, asking what explanations would work better than existing ones. We are asking: What would constitute a better science in this concrete situation? In brief, retroduction means asking why and then developing the answer by thinking and doing research.

S I M E (structural-institutional-mechanism-event) causal theory An approach which acknowledges the holistic part of the social world but also recognises that, once cases which are either unique or comparable units are delineated, we may find structural and institutional causality in how one entity affects another, among these various things (cases and units). For example, the United Nations has a bureaucratic structure and also reflects the power structure among nation-states that are members; at the same time, institutional change takes place within the United Nations and in each nation-state, and mechanisms could be cited that arise from agents' actions, from states' decisions, or from the nature of the relationships in the global structure. In S I M E theory, the events that we record are seen in the context of the whole set of structural features, institutions, and mechanisms of the social and natural world. There are such dynamics that, at the same time, events are not simply caused but also have effects. This causal theory can be represented in many ways: using a diagram of a theory of change; using a statistical model; making a spider diagram or Venn diagram; using NVivo codes on a diagram and so on.

Strategic structuralism An approach to the explanation of social phenomena using a plurality of causes, of which structures play a part, influencing the workings of other mechanisms. The strategic part lies in that the researcher or theorist is carefully picking out particular aspects of society in order to achieve their explicit or tacit purposes while doing a piece of research, and that strategies can be simple (first-order) or complex (second- and third-order, i.e. taking into account others' thoughts and feelings, and our own social influence).

Structures. Also called social structures A structure is a set of parts, all existing in relation to each other in enduringly patterned ways, such that the whole is an entity that has features beyond the characteristics of the parts. These features are known as the emergent properties of that structure.

Sufficiency The capacity of a specific entity to cause some other outcome, by virtue of some essential features. By 'essential' features, however, I do not mean fixed features, but rather those characteristics which have been picked out as key to the nature of the entity, and are intrinsically meant, or invoked, by its name; or which are key to this mechanism. For example soap is sufficient to clean hands (as long as the context is that there is a handwashing action, and water). A further illustration of sufficiency is given by the use of a hand-cleaning gel based on alcohol, which may require neither 'soap' nor 'water' *per se*).

Sufficient statistic For a given model, which represents a real situation, for which we either have data or we know some facts about the kinds of shapes and relationships in the prospective data. A sufficient statistic is an expression which summarises some relationships in the data. If we then add to that expression an additional element, such as shown below, we are adding a nuisance statistic. It is hard to know which element is the nuisance item in a real situation. Under real complexity, it could be that the more complex model is a better representation and achieves more (or a different purpose) than the original model.

Systematic mixed-methods research Methods of research which include tabulated evidence, tables of words, or structured data arising out of any evidence type, including qualitative field experiences or documentary and textual data.

Thresholds For measuring consistency, we want to decide whether to accept a configuration's effect or not, so that the Boolean reduction phase can use that configuration as a positive (causal) term. Therefore we set a minimum consistency level, for example 0.75. Those configurations that reach this level or higher are kept, and Boolean reduction then simplifies a complex expression such as: Thus in the simple example above, if our purpose is to explain Y with low residuals, and M has no correlation with Y, and no correlation with X and Z, d is a nuisance statistic. But if M is correlated with anything else in the model, then for this purpose it is unclear what is nuisance. Finally, if our purpose is to expose as many causes and proxy causes as possible in the underpinnings of Y, then d is not a nuisance statistic, as long as M has some non-zero correlation with another random variable. But, if all the correlations are zero and there is no heteroscedasticity, it is still the case that d is a nuisance statistic. These deductive facts about systems of random variables depend on a grounding assumption of no unobserved variable bias and thus, no relevant omitted factor such as P or T. To make this very clear, if the model did not have a time element, but all these variates are changing over time in diverse ways, then it is less likely that d is a nuisance parameter because the correlations will move around and be non-zero over time.

Venn diagram An image rooted on circles, with a series of cases implied as dots—usually kept out of sight—with dots that have characteristic X inside the X circle, and those without it lying outside of the X circle. The Venn diagram can use squares, as found in TOSMANA software. When Venn diagram circles overlap, the shared area is the 'intersection' reflecting X and Y, and the total area including all of the circles is the 'union', reflecting X or Y.

Index

The manufacturer's authorised representative in the EU is Springer
Nature Customer Service Centre GmbH, Europaplatz 3, 69115 Heidelberg,
Germany. If you have any concerns regarding our products, please
contact ProductSafety@springernature.com

Printed and bound by CPI Group (UK) Ltd, Croydon, CR0 4YY
29/04/2026
02099470-0008